The Future Is Here

Senior Living Reimagined

Lisa M. Cini

THE FUTURE IS HERE
SENIOR LIVING REIMAGINED

iUniverse books may be ordered through booksellers or by contacting:

iUniverse
1663 Liberty Drive
Bloomington, IN 47403
www.iuniverse.com
1-800-Authors (1-800-288-4677)

ISBN: 978-1-4917-8788-5 (sc)
ISBN: 978-1-4917-8789-2 (hc)
ISBN: 978-1-4917-8790-8 (e)

Library of Congress Control Number: 2016900885

Print information available on the last page.

iUniverse rev. date: 02/02/2016

Dedication:

To my family: the past, present and future.

Contents

Foreword

My grandmother, who is 93 years old and is now living with me and my family, has experienced more change in her lifetime than I could ever imagine. Not only has she seen everything that I have seen (which will make your head spin), but she has seen what could only be described as magic, compared to what her grandparents experienced. For those of us blessed enough to have our kids know their great-grandparents, time starts to have meaning. The three largest transformations in their lifetimes have been in plumbing, communication/entertainment and transportation. You realize this when you hear stories about how they met each other and experienced their first ride in a Model T.

My grandparents were raised in homes without utilities/services. Homes were primarily a box composed of walls, floors, windows and a fireplace. They washed clothes on a washboard, then progressed to a paddle washer, then when power came into their homes and the home appliance revolution took hold, they were able to move into gas and electric washers and dryers. They started off with an ice chest and root cellar to prevent their food from spoiling and rain barrels for their water, and progressed to a refrigerator with a frost-free freezer and water that comes out of its door. Candles and gas lamps were replaced with electric switches that would neither burn you nor burn out.

The horse and buggy were replaced with the car, trains and the airplane. Individual mobility that allowed people to go further and not on predetermined routes redefined our communities, from walking distances to distant suburbs, with some living in different states than where they worked. I remember my grandparents telling me how they drove from

Ohio to West Virginia in their friend's Model T so that they could be witnesses for their wedding, and they ended up getting married too! By the time they got back they'd barely thought up a plan to explain their elopement to their immigrant parents. They said that the Model T was so exciting and gave them so much freedom that they forgot about what they would face when they got back home.

However life-changing all of these innovations were, there has been no greater change than in communication and entertainment. My grandparents wrote letters and sent telegrams. They attended live entertainment and read newspapers, books and magazines. Radio, movies, telephone, TV and now the Internet have changed how they get information, and entertainment is delivered in a nonstop global fashion that is deafening at times.

My point is that change will happen, it always does. Seniors are much more prepared for change than most think. They roll with it, adopt what works well and ignore the rest. We could do a lot to make it easier for them. A recent example is when the phone rings and my grandmother tries to answer the remote for the TV … We all laugh, but the devices do look almost identical. To her, because she has seen so much, she accepts the notion that maybe her remote could also be her phone, like my Samsung phone is. When the lights sense her movement and automatically come on, she thinks it's "neat." Technology is getting better and cheaper and has allowed her to stay with me in my home, even though she does need monitoring and assistance.

While Mosaic designs award-winning senior-living homes that look a lot more like country clubs or high-end resorts than they do hospitals or scenes from *One Flew Over the Cuckoo's Nest*, I have chosen to engage in a life experiment with my 93-year-old grandmother and my 76- and 74-year-old parents moving in with us (four generations in the house). It's not only because I want to see if technology can really solve most assisted-living issues, but also because I know that even though my clients love their residents, have the most wonderful environments and provide excellent care, my grandmother would die shortly after moving in simply because of the disconnection with family. The disconnection would not be intentional, but it would happen. Stairs are non-existent and discouraged and her muscle tone would definitely decline. Trying to arrange a Skype session or checking in to see what she really ate is not easily done via technology in senior living, nor does the staffing ratio help. I travel a ton and physical visits would be difficult. She needs to know she is still needed and loved, that she still has purpose, even if it's just saying hi to my 17-year-old daughter when she comes home from school or letting the dog outside.

We have created a senior-friendly apartment, installed sensors for safety, utilized universal design methodologies, provided a heated bidet toilet seat, and created an environment that still makes her climb a couple stairs (five for exercise) to get to the kitchen. While this is far from perfect and only baseline in the technology available for senior living, it is interesting how these simple, little things make a difference in her ability to function independently and enjoy a wonderful quality of life.

As such, I invite you to join me on this journey of understanding the past, present and future of technology and how it will change the face of senior living. I am convinced that the industry will be revolutionized within the next five years with the technology you will learn about in this book. My hope is that you will find something to adopt and improve your current homes, programs and engagement levels, and I can promise you it will be a win-win for all.

The Future Is Here … …
Senior Living Reimagined

My interest is in the future because I am going
to spend the rest of my life there.

Charles F. Kettering

Position yourself to be the leader in technologies
that will change the face of senior living.

CHAPTER 1

The Future Is Here ...

Position yourself to be the leader in technologies
that will change the face of senior living.

1. Are you still selling horses, or are you a Henry Ford? Technology is no longer only for "geeks." It is present in every aspect of our lives.
2. Learn what six technologies will change the face of senior living: sensors and wearable technology, OLED lighting, crowdsourcing and gamification, robots, data mining and artificial intelligence, and 3D printing.
3. Neuroplasticity and the trainable brain.

We are currently caught in a place in time where the apple cart is not only being turned upside down, it is being reimagined. The Internet has created a network of brains not unlike our own that can learn from each other in real time and share breakthroughs that a previous isolated society could never imagine. This change is global and unstoppable. Now we not only have group learning and thinking but we also have the energy and computing power to do this not just within a government or university but worldwide. Imagine Leonardo da Vinci, Michelangelo, Einstein, Tesla, Benjamin Franklin, Edison, Galileo, Newton, Beethoven, Socrates, etc. all being alive today and able to chat in real time to create and solve the world's problems. Add to this Moore's Law of exponential computing power and you start to get a glimpse of what I'm talking about. Now let's go a little further and recognize that pedigree and education are thrown out the window and anyone of any age can participate, not just the chosen few. The simplest point I am trying to make is that technology will change the future face of senior living in the next five years and no one can stop it.

Most of us are completely unaware of advances in technology even though we actively engage in its daily use. Take the automobile for instance; it now embodies 14 of the technologies that I will discuss in this book. For most of us a car is a tool to get us to where we want to go, and provides us with the ability to earn an income, get food and medicine, visit loved ones and have the freedom to explore. However, the car has become so much more than just a transportation tool. Depending on the model and make you buy, a car is now complete with Wi-Fi, sensors, computer systems, memory devices; it is integrated with your personal technology devices like your smart phone; and it has GPS and medical testing (for alcohol). It can entertain, self-park, stop you from rear-ending someone, control your speed, help you navigate, keep your bottom warm or cool, sense if you're falling asleep, and do numerous other things. But with all this technology, it still cannot keep you from getting in a fight with your wife over directions or how you drive.

While we may think the Google car is a far-out idea, in reality it's really very, very close to hitting the masses. Now that self-driving cars are legal in both Las Vegas and California, the genie has been let out of the bottle and will be very difficult to put back. The arguments of safety, freedom and reduced burden on our infrastructure are well documented; now it's a matter of social acceptance and reduction in cost.

Looking back, air travel had the same issues. It was not until the 1950s that air travel was commonplace. In 1937, over a million people took flights, even though travel by train was still the most popular form of travel. After WWII, military planes were used for commercial travel and we have never looked back due to being able to go farther, faster and in more comfort than traveling by train or car. What's missing? And why would Americans even consider riding in a self-driving car? Air and train travel limit your freedom. Traveling by air is now fraught with travel delays and feeling like a convict in a police lineup every time you have to go through airport security. Unfortunately, trains have not kept up in speed or comfort in the U.S. It actually makes perfect sense that Americans would accept self-driving cars for the benefits of freedom, safety (as we have seen with texting) and the ability to relax, work or play, all while getting to your destination.

In practical terms, consumers (your potential senior-living residents) will adopt technology before you (senior-living providers), because studies show that 90 percent of seniors have no desire to enter into senior living and technology exists today (even at Lowe's Home Improvement Store) to allow them to stay at home longer.

Technologies that will make this possible (some of which are available now) are medical self-testing and monitoring, 3D printing of dentures and hearing aids, mobility devices, and communication and safety tools that allow seniors to stay connected with loved ones to ensure they are managing well. Neuroplasticity and brain games, data mining and artificial intelligence will help get the correct diagnosis and treatment faster, which will increase positive outcomes. If you're unsure if Mom has developed dementia or has a urinary tract infection, you can now buy a UTI test kit for $12.99 at the local drug store without a prescription instead of taking a day off work and paying a $25 co-pay, waiting a day for results and then having to do a follow-up exam. If she does have a UTI, then the medicine is $5.99 and you can ensure that it's OK to take with her other meds by talking to the pharmacist. You're able to save time, money and frustration!

Now imagine this with sensor technology that lets you know Mom is taking her medications correctly and has eaten today. Skyping with her on the tablet lets you connect to her more than if she were in assisted living, and seeing her face and her seeing yours and the grandkids is priceless. Mom may even get her hair done and put on makeup because she knows you will see her. Not only can she stay in her home, but she can stay connected. One of the greatest fears of going into senior living is losing the connection with family and friends. It's an effort to visit, and more difficult to communicate just on the phone than it is seeing each other through Skype or Facetime. The costs of a security and sensor package, along with the tablet, Wi-Fi, bidet toilet seat, wearable technology (pendant or watch), and a weekly housekeeper is minimal compared to senior living. If 90 percent of your potential residents don't want to move and now they can get similar services for much less than you provide, which do you think they will choose?

We have seen this all before with disruptive technologies. Where I grew up, in Canton, Ohio, they had a record store chain called Camelot Music – you may remember it. When digital music came out, it killed the need to go out and buy a cassette tape or album. People thought that everyone would want to hold the music and have that pride of ownership, but in reality the joy of the music itself and being able to get it faster, carry your entire collection in the palm of your hand, share it and buy only what you wanted trumped the "pride of ownership." Let's develop this idea further. Now with the Internet everyone's an author, musician, artist or celebrity. The rules of who can be what and where and how services are provided have been stripped away, and all that will be left is what works best for those who are paying. The key is who can bridge the gap and provide solutions that offer value to seniors through technology that will be better than what they can find on their own. Are you up for the challenge? If you are, then you will be ahead of the pack! So let's get started …

Technologies That Will Change The Face Of Senior Living

TECHNOLOGY	SENIOR LIVING								
	Independence	Dignity	Safety	Freedom of Choice	Active Lifestyle	Personalized Care	Quality of Life	Physical Fitness	Mental Fitness
Integrated Data Management Systems			●			●	●	●	●
Off the Shelf Medical Testing	●	●	●	●	●	●	●		
Safety and Sensors	●	●	●	●	●	●	●	●	●
Induction Looping Technology	●	●	●	●	●		●		●
Exoskeletons	●	●	●		●	●	●	●	
LED lighting	●	●	●		●	●	●		
Crowdsourcing	●	●		●	●		●		
Google Car – Legal in Nevada and California	●	●	●	●	●		●		
Robots	●	●	●	●	●	●	●		
Infinte Cloud Computing	●	●	●				●	●	●
Wi-Fi	●	●	●		●	●	●	●	●
Neuroplasticity and Brain Training	●	●			●	●	●		●
Gamification	●	●		●	●	●	●	●	●
Wearable technology	●	●	●		●	●	●	●	
3-D printing	●	●	●	●		●	●		
Drones – Amazon Testing	●	●	●		●	●	●		
Google Glasses Open Public Trials	●		●		●		●		

Since the beginning of time, man's ability to employ the use of tools has differentiated him from the rest of the animal kingdom. Not only can he make tools, but he uses tools to create other tools.

Utilizing rocks, grass and sticks to make spears allowed for more effective and safer hunting. Stringing together willow or other materials into a matrix would trap fish. The first known net dates from 8300 BC; later improvements included rocks being added to the net to help sink the net further. Stone tools dating back 2.6 million years have been found in Ethiopia. There is no dispute that man and tools have a long past and that not only has our success been dependent on this relationship, but that we NEED to create bigger, better, faster, smaller and smaller, stronger and smarter tools. We are entering a new age where we are no longer limited by geographic location, communication, material or cost. The world has been completely connected with past tools that will allow us to create new tools that will make the next 25 years unimaginable.

In order for us to understand where we are headed, we need to agree on some basic definitions for the purposes of discussion. These are not the ultimate definitions of these terms and one could argue there are better ones, but for the purposes of this conversation let's agree to the following:

Technology is: the application of scientific knowledge for practical purposes.

Science is: from Latin *(scientia),* meaning "Knowledge." Modern-day definition is a way of pursuing knowledge.

Tool is: a physical item or procedure or process that can be used to achieve a goal. The knowledge of obtaining, constructing and using tools is technology.

By these definitions, then, pursing knowledge for practical purposes to create tools that help us to meet our goals is what makes us uniquely human. Great Apes are the closest animals to us that use tools; our largest advantage in developing better and better tools was our ability to communicate. Language could be argued as the most instrumental tool to humankind. But there is another unique human characteristic in that we create to create; we experiment when there seems no basis for what the tool could even be used for. We are not smashing nuts open with rocks just to get food, we are building dark matter particle accelerators with the hopes that the invention or "tool" will reveal secrets to us otherwise unknown.

In the creating, we get rid of old technology if it is no longer useful and embrace and adopt the most effective technology we have available. An example of this would be the eight-track music player. Back in the late '70s early '80s, when I thought my parents were so cool, we had a tricked-out van with an eight-track player that pumped out Eric Clapton's "Cocaine" and Lobo's "Me and You and a Dog Named Boo." It was amazing to be able to skip to the song you wanted, versus with a cassette, where guessing was an art form as to where a song started or ended, or what song you were even on. The point is, we adopt new technology often, but the value has to exceed the pain of change.

Now let's open your mind to rethink what technology is and could be. Understand that if in fact these definitions are correct, then technology is not to be feared but embraced, because technology is not about complexity or electronics, it is about creating a tool that will improve your quality of life. Take for instance this photo that I took of a shopping cart in a pharmacy in Germany. While it is low-tech, I know that most seniors in the United States would be raving about this "technology" that not only has a purse hook but also a magnifying glass.

The mere fact that you decided to read this book puts you ahead of most of your competition. You will have the tools to position yourself to be the leader in technologies that will change the face of senior living. By the end of this book you will understand new and future technologies that will transform senior living, and how you can take advantage of this knowledge to reduce risk, increase quality of life and positively impact your bottom line.

> People are afraid of the future, of the unknown. If a man faces up to it, and takes the dare of the future, he can have some control over his destiny. That's an exciting idea to me, better than waiting with everybody else to see what's going to happen.
>
> — **John H. Glenn**

So the elephant in the room is ... fear. Fear of the future.

Although the book *1984* did a great job of instilling fear of sensors and surveillance, Hollywood has taken fear of technology to the next level. Most concepts start off with the idea that technology was created for altruistic ideals to help others, as in the movie *I, Robot,* but the technology gets corrupted either by man or it starts to think for itself and enslaves us.

Tools have always had the ability to perform good or evil. Knives can be used to carve out cancer or to kill; a hammer can be used to drive a nail through a board or hit a thumb, causing immense pain. Tools can be misused by accident or misused on purpose to create harm. History has taught us that using tools for good has always outweighed the bad, so we carry on creating more and more technology in hopes that we'll move higher and higher up Maslow's hierarchy of needs, satisfying our basic physiological needs with technology and now moving on to satisfying our safety and security needs.

Now that we have acknowledged the elephant in the room and understand that fear is a choice, we must move on to understanding where we have been. Confidence for the future is created by understanding our past capabilities and our present circumstances. A tightrope walker does not perform his first walk across Niagara Falls.

Confidence is mental muscle memory of successful experiences of the past. Anyone who has ever learned how to drive a stick-shift car develops muscle memory so that the act seems almost automatic, allowing the driver to move to the next goal or challenge.

While we could discuss the history of technology for an entire book, let's limit it to technologies that have either critically impacted the direction of human history or that we have used from the beginning of time and continue to utilize in some form or another. Our past entails the Agricultural Age, Industrial Revolution and the beginnings of the Information Age. As Glinda the Good Witch says in "*The Wizard of Oz*", "It's always best to start at the beginning." So let's follow Maslow's hierarchy. Physiological needs start with breathing, water, food, etc. Then we move up the ladder to safety: health, security of body, property, work, etc.

Water – we use it to drink, cook with, and clean our food and our bodies. Without water there is no life. Clean water is how we stay healthy. We judge a society based upon their ability to deliver water to their people and provide sanitation systems for waste removal. According to Water.org, 3.4 million people each year die from a water-related disease. Throughout the past, kingdoms were won or lost based upon whether their water supply could outlast their competition.

Natural water sources (springs, streams, rivers, lakes and oceans) were our obvious first choices for access to water. However, as society started to become less migratory, the need for stable water sources not always near a surface water source was developed. Around 10,000 BC, wells were prevalent for access to water but still did not handle sanitation.

Crete, home of the Minoan civilization off the coast of Greece, had an ancient city named Knossos. Knossos was the first civilization to use underground clay pipes for sanitation, aqueducts for water supply and a separate system to handle runoff water. The queen had the first water-flushing toilet, believed to have been built over 4,000 years ago! Not until 1596 was there a water-flushing toilet where the water was part of the toilet.

In the 1830s, sewers systems emptied into our streams and rivers, causing outbreaks of cholera. While today in most developed nations water is clean, sewage is treated and few diseases come from water, even in this world of cell phones everywhere, 2.5 billion people (40 percent of the world's population) don't have access to a toilet.

The infrastructure is the key. It's easier to have a mobile phone than to build indoor plumbing, which requires collaboration, money and equipment. The Bill and Melinda Gates Foundation thinks this is such a big issue that they are challenging us to rethink how we do sanitation. The Bill and Melinda Gates Foundation's Water, Sanitation and Hygiene program focuses on developing innovative approaches and technologies that can lead to radical and sustainable improvements in sanitation in the developing world. They have initiated a contest to "reinvent the toilet" so that systems are self-contained.

So you're probably wondering what this has to do with senior living and technology. Our systems for plumbing were developed thousands of years ago, with the last invention used in most senior-living environments being developed in 1890. That's technology that is 124 years old. Within that timeframe we have been witness to the power airplane, Model T, modern assembly line, phone, TV, FM radio, space travel, fax, 3D, Internet, the microchip, laptop computer, cell phone, GPS, Wi-Fi, LED and so on and so on.

Although I will get to why plumbing is critical to your future success in senior living later on, the case I am creating is that the past is the past. We keep dressing it up but it's still the same old technology. Most of senior living has been dragged kicking and screaming into the use of technology, only adopting new technologies when mandated by law or with fear of a lawsuit. A prime example of this would be the conversion of EMR (electronic medical records) that was mandated to be completed by 2012. It's 2015 and most providers are still transitioning to this technology.

In the next chapter, I will give real-world examples of how companies that recognized and adopted technology saved them and those that chose to ignore technology or hide from it went by the wayside.

CHAPTER 2

Tale of Vision, Fear, Ignorance

To illustrate this better there are three tales I would like to tell you: one of vision, one of fear and one of ignorance.

I grew up in Canton, Ohio, a big small town that in its heyday had massive industry. You may have heard of Republic Steel, Diebold, Timken Roller Bearing, Belden Brick and The Hoover Company. Canton is also known for The Professional Football Hall of Fame and was home to the 25th president of the United States, William McKinley, in 1901. Due to the massive industry, Canton developed several millionaires and was also known as "Little Chicago" thanks to the mob influence. I had this validated by my grandmother, who told me her mother, Assunta DeCosmo, used to yell at the Capone boys to leave their home after they picked up their wine during prohibition. During this time, Canton flourished and had one of the top entertainment centers in the country, called Meyers Lake. It was an amusement park, dance and concert hall with anything from the Glen Miller Band performing to a waterski show. There is no doubt in my mind that money was flowing in Canton for luxury goods and services, having raised the common folk to middle-class citizens with new wants and needs.

Tale of Vision

On a field trip when I was younger, we went to see the home of William Hoover, who was born in 1849. When Mr. Hoover was young he was a tanner in the leather business and built saddles, as the story goes.

Henry Ford started making cars in 1896 and produced the Model T in 1908. Apparently, William saw that cars, although a new technology and expensive, were here to stay. He worried what the market would be like for himself making saddles, as these cars would surely replace the need to use horses for transportation.

During this same time, his wife was loaned a vacuum by a cousin and she raved about it. He had the vision and courage to shift gears and apply his skill set to technology for the home. In 1908, he purchased the patent for the upright vacuum cleaner from the inventor of the vacuum that his wife had tested. Hoover not only acknowledged the future, but took action. This shift allowed him to transition his company through what could have been the death of his business; instead, the company was so successful that he opened offices in Canada and the UK and Australia. Folks in the UK still call it "Hoovering" to this day.

Tale of Fear

The Urban Dictionary states a "Kodak Moment" is a rare, one-time moment that is captured by a picture, or should have been captured by a picture. How then could a 134-year-old company that invented digital technology for the camera in 1975 go bankrupt? People are taking more pictures of themselves, others and things than ever before. We are living in an ultra-documented world. The answer is fear. At the time, 95 percent of Kodak's revenue was from paper and chemicals. Although they had invented the technology that would have been their savior, they were too fearful of losing their current cash cow; this paralyzed them from moving forward. Fear is an interesting concept; it skews how we perceive reality. They forgot they were capturing a rare, once-in-a-lifetime moment; they thought they were selling paper and chemicals. Their legacy is now plagued by headlines of "Kodak's inability to evolve led to its demise." RIP Kodak, 1878–2012.

It is a contrary uplifting tale that in 2010 two guys created Instagram, a free app to share digital photos. Instagram was acquired by Facebook in April 2012 for one billion in cash and stock, with 13 employees. Instagram understood that their mission was to share moments. Once the masses decide to adopt new technology there is no stopping it; it goes viral and you either need to be part of the wave or will be buried under it. Kodak could not have only been part of the wave, but ridden it; instead, they fought the wave and tumbled upside down and sideways, losing all sense of direction until their demise.

Tale of Ignorance

In 1960, Sabre created an online booking program for travel agents, born out of a conversation several years earlier between the president of American Airlines and a senior sales rep of IBM. They sat on a plane in 1953 and discussed how data management could make reservations real-time for any travel agent anywhere with a computer. Travel agencies partnered with the airlines, which paid them a fee for every ticket booked, and the system worked well. Somewhere between 1980–1989, Sabre started offering this service direct to any consumer with a personal computer. As consumers became more tech-savvy and the database management systems became more user-friendly, it opened up the market to cut out the travel agent altogether and go directly to the consumer.

The travel agent industry overvalued their market proposition, and could not see that lower-level tasks such as searching for a flight and booking it were seen as holding little value by the customer. Customers needed a travel agent if they needed expert advice on traveling to a location they had not been to or on where to stay when they were unfamiliar with the town, but a flight was like a bus or train ride – it was simple "get me from A to B on this day." The transaction was very similar to the old system of telephone operators that would connect you to another caller. When phone numbers were able to be dialed directly, this entire system was eliminated to only having "0" being called when there was an issue. I am not even sure if dialing "0" works today or what you would use it for! Finally, in 1996 Travelocity (previously the Sabre system) came out of the closet and never looked back.

Customers had spoken. Technology made the lower-level tasks of making flight arrangements easier and quicker and with less frustration than explaining what you were looking for to an agent and then waiting for them to respond. Movies, books and shows make jokes that if you see a storefront travel agency it's a front for an illegal business because they are no longer a viable business. Ignorance not only of their value to customers but also ignorance of how quickly consumers would adapt to technology resulted in 61 percent of travel agencies closing down since the 1990's due to their lack of being able to see the writing on the wall and shift gears. The ones that have survived have entered specialty markets of luxury and business travel.

Hoover saw the horse for what it was and shifted his skill set to a new technology. Kodak allowed fear of losing current revenue to cloud their mission and vision of the future. They did this even though they created digital film technology prior to their competition and should have been the leader in the industry. And finally, U.S. travel agencies were ignorant to the fact that their clients had access to the same technology they had been using and now they had become a barrier to getting tasks completed cheaper, faster and with less frustration.

So are you still selling horses or are you the Henry Ford of your industry?

While there are numerous other industries that have felt this same burn (take, for example printing, with the ink-jet printer), it's not too late to recognize what is coming down the pike and how best to respond to it. Knowledge is power. It allows you the ability to develop a strategy. To others you will seem visionary, but you will have done your homework and leveraged your knowledge of the future to obtain the best quality of life for your residents, reduce risk and improve your bottom line.

It's an exciting place to be, right here, right now, leveraging the past and the present and peeking into the future.

To understand the past and the present you need to understand Moore's Law and exponential growth (we will discuss this in later chapters). Moore's Law basically states that computing power doubles every 24 months. Peter Diamandis presents the following data in his book *Abundance*: "In 1971, the Intel 4004 integrated circuit had 2,300 transistors costing $1 each with .00074GHz of processing power. Today, an Nvidia GPU has over 7.1 billion transistors, which cost less than $0.0000001 each and has over 7GHz of processing power. This is 10,000 times faster, 10 million times cheaper, and 100 billion-fold improvement." This explains how computers can get smaller and smaller yet be faster and smarter. It is said that a kid in Africa with a smart phone has more technology than President Bill Clinton had at his disposal when in office. Everything we use that has a computer in it has been impacted by Moore's Law.

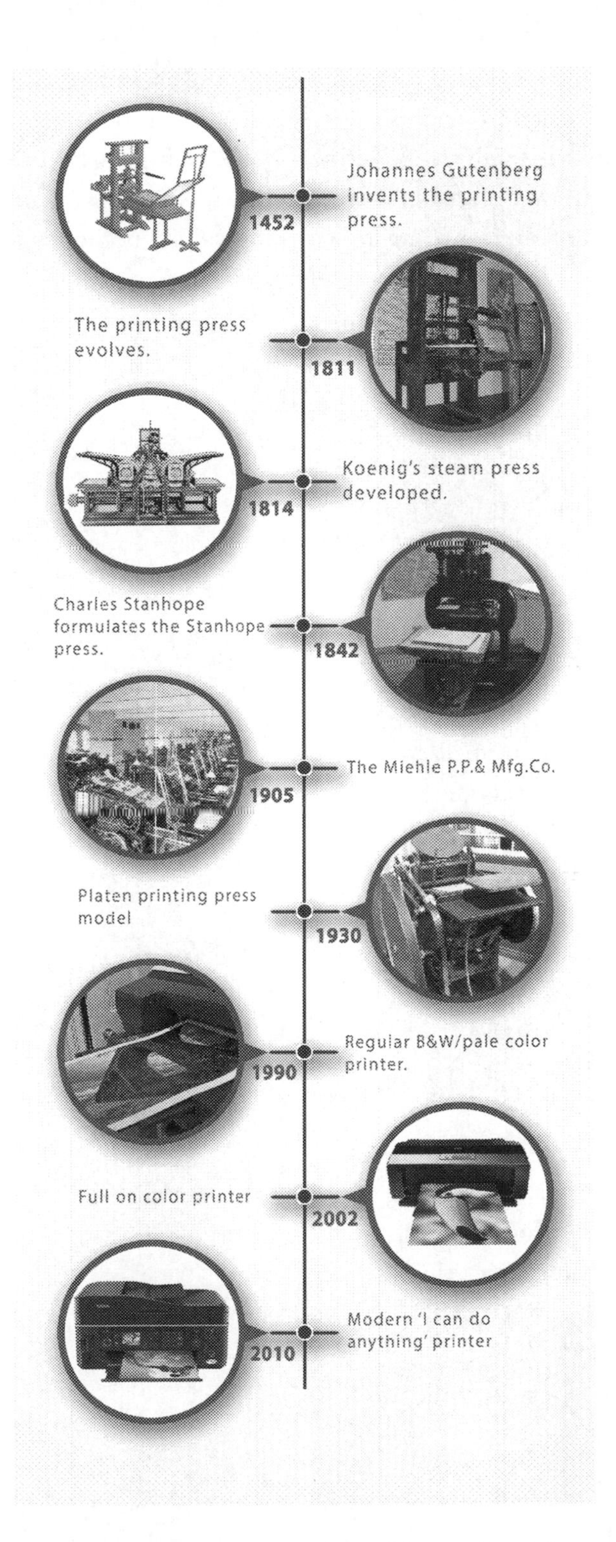
1452
Johannes Gutenberg invents the printing press.
The printing press evolves.
1811
1814
Koenig's steam press developed.
Charles Stanhope formulates the Stanhope press.
1842
1905
The Miehle P.P.& Mfg.Co.
Platen printing press model
1930
1990
Regular B&W/pale color printer.
Full on color printer
2002
2010
Modern 'I can do anything' printer

Present technologies express Moore's Law. Take a moment and make a list, either mentally or on a piece of paper, of all of the technology you currently use on a daily basis. Given the definition that technology is the application of scientific knowledge for practical purposes, what do you come up with? In a couple of seconds, I come up with over 44 technologies that I use on a daily basis, and I did not include software or apps. If I had, I am sure the list would have been twice as long.

1. Phone
2. Cell phone
3. Car
4. Electric curling iron
5. Sonic toothbrush
6. Hair dryer
7. Indoor plumbing
8. Alarm system
9. GPS
10. Cloud computing
11. Laptop computer
12. Tablet
13. Digital camera
14. Fax
15. Copier
16. Vacuum sweeper
17. Clothes washer
18. Clothes dryer
19. Dishwasher
20. Refrigerator
21. Electric auto-dimming lighting
22. Heated flooring
23. Thermostats to electric heat
24. Gas hot water tank
25. Blender
26. Juicer
27. Stove
28. Oven

29. Microwave
30. Coffee pot
31. Food Processor
32. Electric bed
33. Massage chair
34. Scooter
35. Personal printer
36. Glasses
37. Electric iron
38. Safety cameras on car
39. Heated electric bidet/toilet
40. Wi-Fi
41. TV
42. DVR/DVD
43. Outlook calendar
44. Wi-Fi speakers

These technologies are average to most consumers and some would even say they are not necessary. TV reality shows challenge this thought process with hit shows like *Survivor*, *Naked and Afraid* and *Man vs. Wild*. It seems we've become so pampered that it has now become interesting to watch people without technology on TV as entertainment. What we used to call "camping" when I was growing up is now considered a survival test.

Technologies that we are dealing with in the present that should be leveraged by the senior-living industry are those that impact the quality of life of a resident, risk management, and/or items that will improve the bottom line. You may have heard of some of these technologies or even be on the forefront of these technologies in your homes. If so, you are way ahead of the curve.

In designing for thousands of homes over the past 20 years and visiting many more, I am still shocked by the lack of technologies used to make the environment better, cheaper and more connected. It seems there is a distrust of technology that being connected or watched will open a potential for lawsuits and make it even harder to provide care for seniors and stay profitable. Dignity is the other factor that gets brought up as a stumbling block to adopting new technology in senior living. The key is appropriate permission-based use of technology. Great technology is hardly noticeable, as you will see, and it solves problems that improve dignity, safety, outcomes, compliance and risk.

The easiest way to approach this discussion will have to be twofold. First, you need to define your categories of technology that you will need, and then second, these technologies will need to be married to the impact they have on your mission. Basically, the cost of the technology has to outweigh the pain of doing things the old way.

Most senior-living homes do not start their mission with: We strive to make money, keep ourselves from getting sued by keeping the residents from getting hurt, tell the sons and daughters of our residents the least amount of information for fear that they will not agree and call the Ombudsman, and finally, we can't admit to ever making a profit for fear the unions will move in on our staff and turn this into *Attica*.

The typical senior-living mission is something like: We believe in dignity, respect, independence, and acknowledging the individual. We care; we're compassionate and nurturing. We believe in freedom of choice, staying connected to the community, family and friends and staying active. We believe in personalized care and continuous improvement.

Typically, technology is first adapted by those with the lowest mental barriers to using it. Tweens and teens usually fit the bill. When I was having trouble doing research on the Internet in 1998 I asked the 9-year-old neighbor kid to help. He told me to type in the word "Google". I said, "Google? What is it?" He told me, just do it and stop asking questions.

The two missions could not be further from each other, but sadly they tend to coexist due to a lack of utilizing available technology, either out of fear or ignorance.

Technology does require an investment and this is usually the first and largest objection. But I ask you, can you imagine using an outhouse versus indoor plumbing? Having a pail water well to get your water vs. turning the lever on your sink? It sounds ridiculous to us now. The amount of time wasted in villages without running water keeps girls from being educated, not to mention the safety issues involved. Can you imagine going back to a non-electric typewriter? I can't comprehend how long and how many rewrites it would take me to write what you're reading.

Or let's go a step further. Let's say you have access to the Internet but you choose not to use the technology because it's silly, with all the Facebook stuff and the costs to have it. You need to research something so you go to the library (yes, they still exist), and you search through the system to find out which books may be of value. You then get their Dewey Decimal numbers and you're off on a hunt. The books are all over the library and intermixed. They're heavy, but that means value, right? You then look at each one, looking for a hidden treasure of evidence; finally, you find your nuggets. When the information is short, you hand-copy it down but when it's large you mark the page and then copy it at the copier. After all the research is completed you have to return the books. If you missed something and can't remember which book it came from, you have to start all over. There is no cookie history to lead you back to where you were. This was exhausting to write; just detailing it out felt like a waste of time, yet systems in senior living are very much like what I have just described, using archaic systems because they exist and are familiar.

Evaluating the right-fit technologies for you and your residents and then ensuring a return on investment will be the key for those wanting to lead in senior living. Technology applied just to have a marketing advantage rarely pans out. The technology must be aligned with your mission and core values. Integration needs to be embraced from all sides, not just management. Selling to your team will be just as important as selling to your residents and their families.

Currently technologies available:

- Integrated data management systems
- Off-the-shelf medical testing
- Iris – safety and sensors
- 3D printing
- Google Car – legal in Nevada and California
- Exoskeletons
- Induction looping technology
- LED lighting
- Crowdsourcing
- Sensors
- Robots
- Cloud computing
- Wi-Fi
- Neuroplasticity and brain training – Lumosity
- Gamification – Wii Fit
- Wearable technology – Nike
- Drones – Amazon Testing
- Google Glass – open public trials

Technology can help you to increase the quality of life of your residents, reduce your risk, and improve your bottom line.

I believe there are six technologies that will change senior living and everyone's lives.

The question is … which ones will you embrace and adopt?

1. Sensors and wearable technology
2. OLED – lighting technology
3. Crowdsourcing and gamification
4. Robots and technology
5. Data mining and artificial intelligence
6. 3D printing technology

CHAPTER 3

Sensors & Wearable Technology

Infrastructure limits technology adoption, adaptation and access (AAA). Infrastructure AAA, in this sense, is your worst enemy. In the early days of my career, even in Skilled Nursing, oxygen was hard-plumbed into the wall in a patient room, and so was nurse call. If you had a patient who needed oxygen, your options on a new building were to pay the excess cost of plumbing oxygen into every room to ensure flexibility, or have only certain rooms with oxygen and reject a patient if you did not have what they needed. Now, with portable oxygen machines, we don't even use it in a wall if someone needs it in a nursing home. Utilizing portable machines has allowed an estimated 2.5 million seniors in the U.S. alone, who would otherwise need senior living, to stay in their homes.

Now imagine nurse call in senior living, or more appropriately, emergency call. In the past, these were hardwired into the wall, which meant that you'd better have the placement of the bed exact, and guess where the resident may fall and hope that the cord can be reached. Now, emergency call is wireless and you can add more if you need to. The benefit to the resident is that they can arrange their furniture any way they like, and the emergency call button can now be easily relocated.

Remember the old payphones and drinking fountains you had to have in your senior living home? These have gone by the wayside and very soon the hardwired resident room phone will disappear due to the easy access and low cost of cell phones. Cell phones are now priced so low they can cost about the same per month as two gallons of gas for your car. With Wi-Fi you can even call for free internationally with Apps like Viber. The point is, technology has finally broken through the infrastructure barrier. The best way to explain this is to imagine if all of our transportation was still restricted to trains. You could only go where the tracks were, and when and where the train was stopping. Now imagine your car, and it's not just a car but it's one of those off-road cars you see on the commercials that allows you to go over boulders, through creeks and still drive on streets. The freedom is endless as to where you can go, at what time and what you want to see along the way. A whole new world is opened up for you to discover and possibly create better, faster, cheaper solutions. A mode of transportation without rails gives you complete control versus being bound by what the Train Barons set as the schedule, path and stops. Technology is now free from infrastructure. From trains to cars, phones to cell phones, cable to Wi-Fi, nurse call to wireless nurse call, plumbed oxygen to O2-making machines, sensors that can transmit via chemical reaction, and wired electric versus batteries – it's all portable, chargeable, and wireless, which gives us freedom to make choices. Your future residents have made these choices and are often already using them.

Studies cite that 90 percent of your target market wants to stay in their homes, because they don't want to move into senior housing. They will utilize all the available infrastructure and free technology out there in order to stay in their homes longer, unless you offer a better solution that not only embraces their need for connection and freedom, but makes it their right. The key is providing them better and better technology that helps them to be more independent, easier than they can do for themselves at a lower cost. If your current value proposition cannot do this, you have little value to your target customers.

Hopefully at this point I've adequately articulated that this dilemma senior living is facing is nothing new, and is faced by every industry at one point or another. It's how an industry decides to react to the change that is key. Progress will not be stopped: it is like a river and it will find a way to get downstream via the path of least resistance. Allow yourself to discover the possibilities with an open mind and learn what will best work for you so that you can be part of solution.

We don't even realize it but we already live in a world of sensors. The canary in the coal mine was a sensor, letting the miners know when it was time to get out of the mine for safety's sake. The first cars were without sensors; imagine no gas gauges or speedometers. Folks soon realized it was very helpful to know when you were going to run out of gas prior to doing so, when you needed oil, how fast you were going, and so on. I remember my first car was always on empty; it wasn't really, but the gas gauge was broken. I would have to calculate what I thought was my gas mileage constantly, and I was scared to drive on the highway for fear I had miscalculated and I would be stuck. A gas gauge is such a basic thing, yet so valuable. Basic sensors are used on our hot water tanks, heating and air-conditioning units in the form of a thermostat, and temperature controls for the oven. Some sensors are for convenience, others for safety, and some to save energy. Safety sensors alert you to a fire, break-in or a tornado coming; energy sensors may turn off the lights after no movement for a period of time. Sensors are part of our everyday life in one form or another.

One of the simplest, most cost-effective sensor technologies on the market rarely gets discussed and I have yet to see a senior living home have them installed. I bet if I gave you hints, it would not even cross your mind. I first discovered this technology in Japan 10 years ago. While visiting a public restroom I was astonished to experience a heated toilet seat that had a nightlight, was a bidet and blow dryer, and was anti-microbial and anti-bacterial. This was a public restroom! In the U.S. I am happy to see a toilet seat and toilet paper. While this story may seem crude to you, it should not: toileting in senior living is not only an issue for staff but also for the residents' dignity and independence. As soon as I got back home I ordered a Japanese-style bidet toilet seat with all the bells and whistles for my master bath.

A weird phenomenon started to happen at my home ... no matter how many available bathrooms there were in my house, my children's friends, our family and even guests would find themselves drawn like a magnet to my bathroom. At first I thought it had to be my incredible taste ☺ or maybe the heated floor, but on further investigation it was discovered to be the toilet seat!

When we moved to a larger house to be able to have my parents and grandmother move in with us, my only requirement was that we would have these toilet seats on every toilet so that I could have some privacy in my master bath. I have to admit I was also performing my own senior living experiment. I have been pitching these seats to clients for 10 years with no luck and wanted to see firsthand how my parents and grandmother with early stage Alzheimer's would respond. I could not imagine having a technology available to my family to ensure their dignity and independence and not use it.

The results are in: my 93-year-old grandmother did not know what it was until it was explained to her, but now she loves it and also uses it as a shower for her bottom, as she puts it, "in between showers." My parents, at 74 and 76, think it's the greatest thing since sliced bread and feel the need to tell all the family and their friends about it.

Who knew for less than $300 per toilet I could make everyone so happy, increase their independence and dignity, and increase the chances of me not having to assist in toileting in my own home? It's a real win-win. My recommendation would be to trial these in independent or assisted living first. They require no infrastructure and fit on top of an existing toilet. You simply hook the water up and then attach the electric (which your bathroom should already have). The seat tends to be higher than a normal toilet seat, which is perfect for a senior, and with the nightlight you reduce risk and increase your chances of a better hit ratio with your male residents.

As a note, there are many different manufacturers and styles of these toilet seats. I have been trialing five different ones. Most allow you to control the seat and water temperature and then the intensity of the water, direction and the intensity of the blow dryer. My recommendation would be to get the toilet seat that allows the controls to be mounted on the wall so that they don't accidently get used to steady a resident. One final note is that they are pressure-sensitive, meaning they will not spray unless someone it sitting on the seat to avoid someone pushing the controls and water spraying everywhere. These have been tried and tested in other countries and have stood the test of public restrooms – I think it's high time we give them a shot in senior living.

Recently the public was introduced to sensors integrated with the home environment through their iPhones. Users can integrate with Nest to turn the heat up or down before they get home, lock doors and even view who was at their door right on their phone or iPad. You could get a notification that your kids are home from school and they locked the house back up and are safe and sound. The technology centers on peace of mind and has proven to have been what consumers were looking for.

Lowe's Home Improvement smart home monitoring system named Iris has taken this a step further by integrating a system that can be used with any smart phone and has a low monthly fee. Iris has kits that are divided into Safe and Secure, Comfort and Control, and the Smart Kit. Care, which is an add-on to the system, is specifically for seniors in their homes. The system will allow for you to set rules and let you know if Mom has not been in the refrigerator in the past 24 hours (or whatever time span you choose), which may indicate she's not doing well. Cameras can be used inside the home and operated remotely. A notification can be sent to you if the front door opens in the middle of the night. The system is intended to provide greater independence and dignity, but also keep the children from worrying when everything is fine through the sensors and knowing when the four-hour drive is necessary. Both Iris and the iPhone technology are training families to expect more. They know what technology can do and expect senior living to have more than they have access to. Can your senior living environment do this? Be prepared to start having to answer the question.

There are a couple of senior living companies that have great sensor technology, such as WellAware System, QuietCare, GrandCare and Stealth Health. Sensors can aide in documenting patterns and then when the patterns – such as sleep, going to the bathroom, etc. – have changed, the reasons can be investigated at the start of the downward spiral. Having this information could result in re-adjusting medications or early identification of a UTI or early stage dementia. The benefits are enormous.

Wearable Technology

Most of the current senior living wearable tech is in its early stages, or in the form of old-school medical alert pendants or wander-guard technology. The type of wearable tech that we will be discussing is the tech that seniors have access to today off the shelf, and the level of information it provides. Whichever company figures out a way to integrate the current systems into patient charting and staff response times will be the winner in the wearable tech game.

If your reading this book it's a safe bet you're a technology fan or familiar with technology. Fitbit is a name-brand data-gathering bracelet that tracks your activities and encourages you through gamification (we will go into detail on gamification in a later chapter). They capture the hearts of fitness enthusiasts that desire to get better results through tracking and tweaking. As the old saying goes, "What gets measured, gets improved."

Basis is a fitness-and-sleep-tracker watch that was released in past few years but claims to be the most comprehensive health watch on the market. The watch not only documents when you sleep but also tracks your REM and how many times you toss and turn and wake up. It employs gamification and also tracks heart rate, steps, calories burned, perspiration and body temperature. I was given one of these watches and was mortified to find out that I burned almost the same amount of calories while I was sleeping as when I was awake! I guess sitting at your desk all day doesn't get the heart pumping. Just by learning this information, I have become more active about getting up and taking breaks, stretching and taking the stairs instead of the elevator. By all accounts I look fit, but my sleep score was terrible.

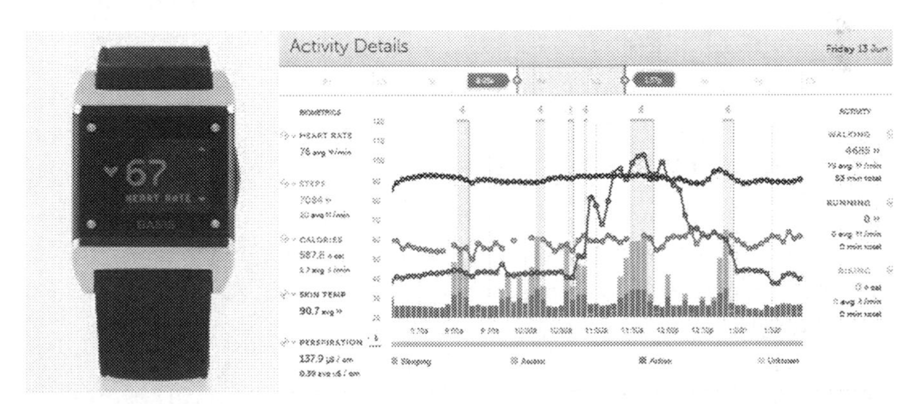

Many experts believe that Americans' terrible sleep habits are at the center of our health issues. The data you can see above is my personal data from the watch that gets synched with my computer and the Mybasis site. For less than $200 for the watch (no monthly fee) I have more health information reported 24/7 than my grandmother would be able to have in a senior living home. As we have discussed prior, it's not the event that we should all be concerned about, but how to monitor the data we can easily obtain to reduce the risk of an event or put in place an action plan prior to the crisis happening.

Hearing aides are a wearable technology that has been around since the seventeenth century, with the first electronic hearing aide in 1898. Cochlear implants became revolutionary for those with extreme hearing loss but required surgery. Now there's a hearing aide from Sonitus called Soundbite that is bone-conducted but requires no surgery and is FDA-approved. The really exciting part of this new technology is they are moving towards the ability to implant a hearing aide in a molar. Imagine a world where you're not having to look for the lost hearing aide anymore.

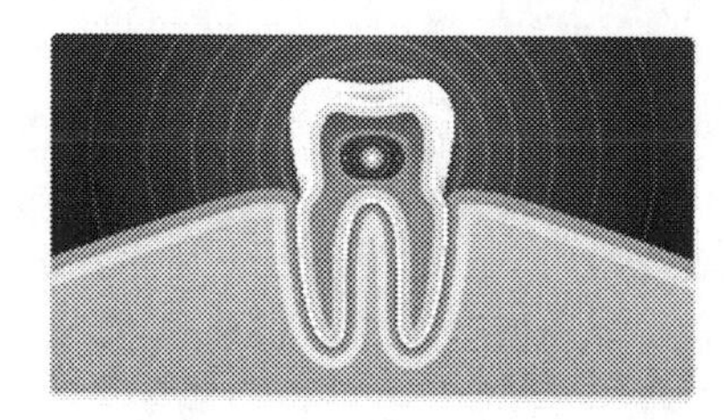

While Google Glass has been in the news over and over again, they have never taken off like Google thought they would. They have, however, inspired others to make glasses that are a combination between the sci-fi Oculus Rift virtual-reality gaming glasses and the glasses actor LeVar Burton wore on *Star Trek: The Next Generation* which allowed him to see even though he was born blind.

What if they could make the blind see, or at the very least, help those that are sight-impaired with degenerative eye diseases see? Imagine how this could help seniors.

Professor Kohn of the University of California has developed such a device. He is using computers and math to correct the distortions seen by those with macular degeneration. He is working on both glasses and contact lenses that would be tailored toward the individual user.

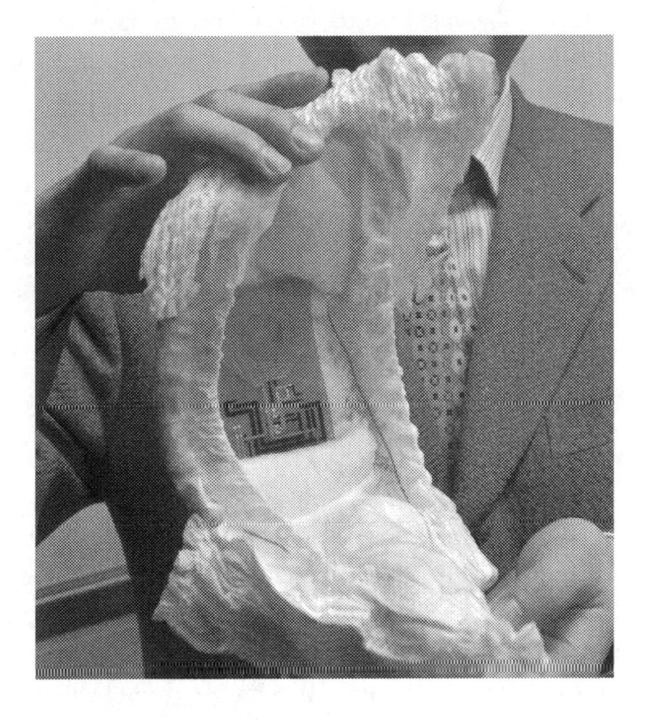

The Japanese have been using smart diaper technologies for babies and these now have moved to the senior market. Adult smart disposable diapers are outselling baby smart diapers due to the rising age of seniors in Japan. The technology sends a wireless signal to the caregiver that lets them know if the diaper is wet. This is a tremendous advancement with respect to resident dignity, especially for those with dementia or Alzheimer's, as the resident no longer has to be undressed to check to see if there is an actual need for a change.

Per Mobi Health News by: **Aditi Pai** | Feb 10, 2014 "Another company, **Pixie Scientific**, which makes sensor laden diapers, launched a pilot study for babies at UCSF Benioff Children's Hospital last fall, and around the same time the company also initiated a pilot study for adults at several assisted living facilities with its new product, Pixie Briefs."

So far the sensors that have been developed in Tokyo measure moisture so caregivers know when to change the patient, but Pixie Scientific's diapers, for both adults and babies, also alert the caregivers when the sensors detect a urinary tract infection, prolonged dehydration, or developing kidney problems."

As you can see, sensors are just starting to surface in everything from how we take our medicine, when to change a diaper, reading the blood glucose levels in our tears, knowing how long we slept, how many times we tossed and turned, and what our sleep quality was, to sensing the electrical impulses in our brain and allowing us to move body parts that are paralyzed.

Athletes and the military are at the forefront of sensor technology, but make no mistake, it is weaving into all of our everyday lives. Soccer shoes have been equipped with sensors to discover speed and how many miles a player runs in an average game. The results are staggering: a pro soccer player runs eight miles during a 90-minute match. Basketballs, soccer balls, and tennis racquets are now equipped with sensor technology to help you understand how to strike the ball better or how often you're hitting the sweet spot with your racquet. Fabrics are currently being released that have sensors woven into them. Our children and grandchildren will not ever argue with the TV over whether or not the ball went in the goal and the referee made a bad call because the ball and the goal plane will both have sensors and automatically signal when a goal has happened.

A quick search on the Internet reveals new categories of wearable tech every day, and health management is looking to be the largest sector. Senior living providers not only need to follow the home integration and safety sensor technology but also follow health monitoring that allows with ease the monitoring of your parents' heart rate and sleep. The game is certainly changing. Now that Apple, Lowe's Home Improvement, GE and others are getting into managing your home connection and security to ensure a loved one is OK and they remembered to take their dinner out of the oven, it won't be long till we all have a "robot" in our homes.

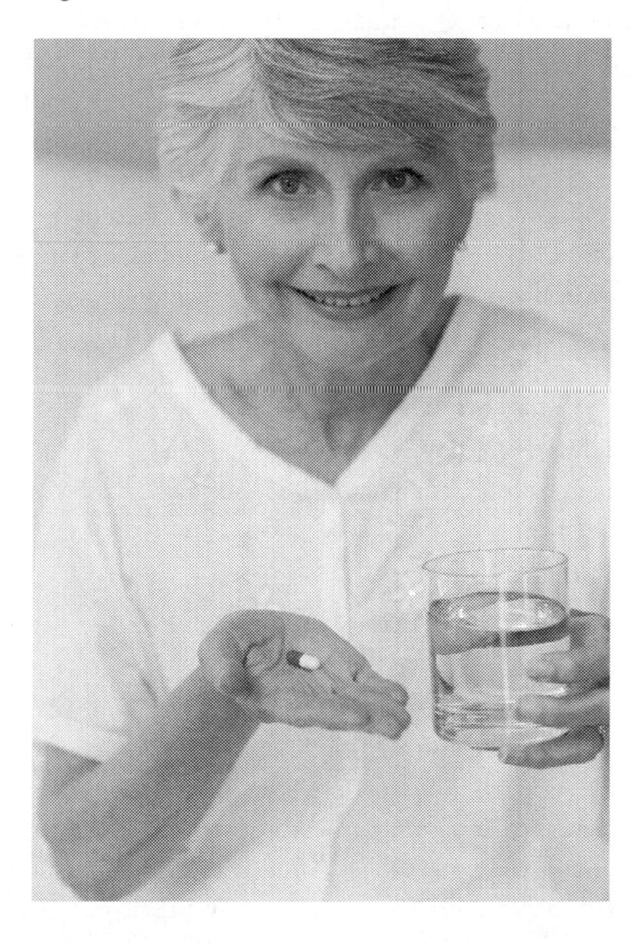

Anyone involved in healthcare knows that medication compliance is difficult. Besides toileting issues, help with medication seems to be a huge driver to getting mom into assisted living after she forgot to take her pills or took too many, sending her to the hospital. While senior living has done a great job at managing medications, it is still unclear if they are working properly for the individual. Protus Digital Health has a solution for this which, when taken to the next level, could also be used for mom to stay in her home longer as data will be immediately sent to the doctor or caregiver via Bluetooth if she has missed her dose or taken too much.

Protus Digital Health has created the first FDA-approved ingestible sensor for medication compliance and feedback. The sensor is about the size of a grain of sand and is mostly made of silicon. The sensor reacts with the stomach acid and creates a type of potato battery with its small amounts of magnesium and copper. This allows data to be sent out to a smartphone or other device. The device passes out through the normal process.

For the caregivers and doctors to know if their resident's medications are actually working could increase positive outcomes immediately, as opposed to by trial and error. Worst case, you would know if medication compliance is even taking place. Medication compliance and feedback could change how prescriptions are managed in the very near term.

Currently Helius (Protus Digital Health's name for the pill) is being utilized for heart failure and hypertension, but looks to add other chronic disease management.

Integrated Data Management Systems

A huge disconnect currently is not having an integrated system that not only has EMR but also risk management, mining of patient data, billing, food service, staffing, payroll, activities, marketing and property management. This creates information silos with these departments which is extremely ineffective and costly. Currently there are nine providers that provide the majority of modules integrated into their system. When you can stop looking for data and let the data work for you, you have essentially cut out the middleman, just as Travelocity was able to. Cheers to those of you who have invested in an integrated data management system! It's a huge piece of the technology discussion, but if it's the only part you have, you still have work to do and the coming chapters will give you the information you need in order to move forward in the best direction.

The Assisted Living Federation of America has an excellent resource for data management systems. There is no perfect solution at the moment, which increases the risk that well-meaning staff will use work-arounds or shadow IT to get through frustrating, inefficient or nonexistent software applications. These can even be on their own smart phone or computer.

The U.S. Department of Human Services handed down in May 2014 the largest fine in U.S. history to two New York hospitals for failing to protect patient data under the HIPPA law of 1996. The fine was $4.8 million and the primary issue was found to be a work-around by a doctor.

The risks have gotten higher with the U.S. government's enforcement of HIPPA and the Healthcare Act. Data management systems can be the key to reducing risk, saving time and measuring outcomes which in turn would lead to better results.

When it comes to sensor technology in senior living, the Germans are at the forefront. I had the pleasure of interviewing Christl Lauterbach, managing director of Future-Shape GmbH out of Höhenkirchen-Siegertsbrunn, Germany, to better understand why Germany has taken to technolgy to help with their growing senior population.

Future-Shape has developed products that help to detect movement on the floor. This is really not all that earth-shattering when you think about it; we use this technology every time we walk into a Walmart or Target. As we approach the door, the door senses us through the floor with our weight and opens up for us automatically as if it were magic. Future-Shapes SensFloor and SensSeat are unique in that they can not only sense weight or steps, but through complex computing, they know the difference between my mother and an intruder.

While the product has yet to enter the U.S., SensFloor is looking to pilot it soon in several senior living locations.

The product was designed to help seniors stay independant longer. In order to do this, ensuring a senior's saftey was paramount and so was understanding that when falls happen, the senior may not be able to call for help if they are unconcisious or unable to get to their pendant or a phone.

When asked what are the benefits for installing this flooring into a senior living home with multiple residents, Christl listed the following as benefits of SensFloor: "Fall detection, fall prevention through activity monitoring, intrusion alarm, switching orientation light automatically to avoid accidents in the dark, intelligent control of automatic doors – keeping doors shut that should stay shut or not opening them if someone walks parallel to the door or merely stands in front of it. Intelligent opening of doors also helps for those in wheelchairs."

When asked if SensFloor would be cost-effective to install in a private smart home for senior safety, Christl responded, "Yes, we just had the opening of a sample house – there will be 100 of these houses in the Vita D'oro senior residence. The installation of the SensFloor system there is approximately 3 percent of the price of the complete house and provides a whole bunch of comfort, safety and security applications."

According to Christl, the reason that Germany has engaged so greatly in technology solutions for seniors is due to their demographic change. "In Germany we had 2.3 million people in need of care in 2009; 710,000 of these are in nursing homes. This number will increase to 4.5 million until 2050. The costs for living in a nursing home is extremely high and a heavy load for the social systems. And most people want to stay at home as long as possible. Just another number how important it is to avoid falls: In Germany we have 100,000 hip fractures per year. The medical costs per treatment are approx. 20,000 euros, which adds up to 2 billion euros per year. According to the statistics, every tenth fall leads to medical treatment – for older people, it's often a dramatic impact to their wellbeing in future."

When asked what the difference is with SensFloor compared to current products on the market and if there were studies to back up any claims, Christl responded, "The studies show that falls are prevented by using proximity sensitive mats for patients that are in danger of a fall, when they get out of bed alone. Especially at nighttime, if the nurse gets a signal when someone starts getting out of the bed, she usually is available for helping. The advantage of our system is that the mats are proximity sensitive and trigger an alarm before the patient stands on their feet."

Pricing is always a concern, and although the pricing is for Europe, currently it equates to 200 euros per square meter of floor area for the installed SensFloor system, including project planning, installation of the SensFloor underlay and the interconnection to the home automation system. The SensSeat is intended for the seats to alert staff when someone is rising from their seat who may need assistance. But they also have used them in research projects with universities, where SensSeat is used for the captain's seat on cruise ships to show that he is sitting in the control seat. The cost for a complete set – a sensor mat, a power supply and a transceiver, giving the alarm (private home) or triggering an alarm for an indoor call system (nursery homes, hospitals) – is around 660 euros.

While we probably won't see SensFloor mainstream in the U.S. for a couple of years, the technology is available to truly make a smart home for a senior. Add in the robotics developments and 3D printed food that Germany is creating and U.S. senior living could do a lot to take their cues from Europe on what technology will or won't work for seniors and investors.

Christl and Future-Shape were kind enough to provide content for their VITA D'ORO project in Germany.

SensFloor® for Comfort and Safety

We all want to remain fit and healthy as long as possible. If we require care when we reach a high age, we at least want to be able to stay in our homes.

SensFloor supports this: Orientation lights that turn on automatically at night prevent falls. When someone is lying on the floor, a fall alarm is triggered and transmitted to the service centre at the Clubhaus. Combined with the burglar-alarm, it is possible to detect unauthorized people entering a building. Movements by residents on the other hand do not trigger any alarms. When a person is living alone, too long a period of inactivity may give cause for concern. In this case the floor notifies the service centre and thus ensures that a person does not have to suffer too much time helplessly. If at some point a resident requires a wheelchair, doors can be activated automatically and intelligently, already opening when someone approaches them. This prevents having to align wheelchairs with them. Doors remain closed if someone merely stands in front of them or passes by.

SensFloor can be combined with a protective system that acts to prevent residents suffering from dementia or prone to running away from leaving unnoticed. All these functions can individually be added at a later date without any reconstruction work necessary.

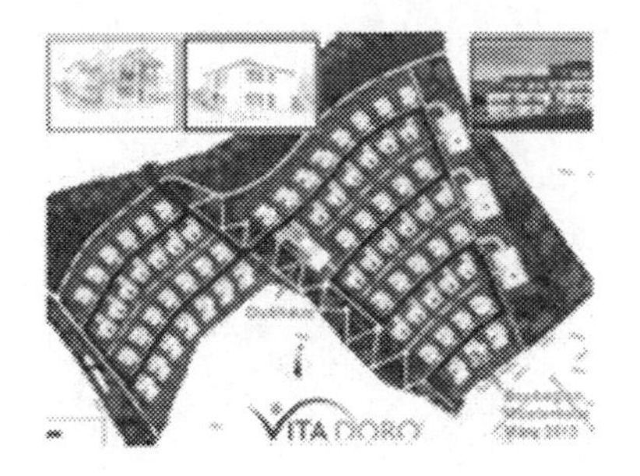

The core of the residential area is the VITA D'ORO Clubhaus (clubhouse). It is equipped with a swimming pool, multi-purpose rooms, a lounge, a library, therapeutic rooms, and a sauna area, as well as a technical centre with a 24h on-call emergency service for the entire park. This is also where all alarms from any buildings are transmitted to.

The SensFloor underlay is installed in the apartments and buildings beneath TILO parquet, DLW linoleum or carpets. It is entirely invisible and serves as a buffer against footfall noise.

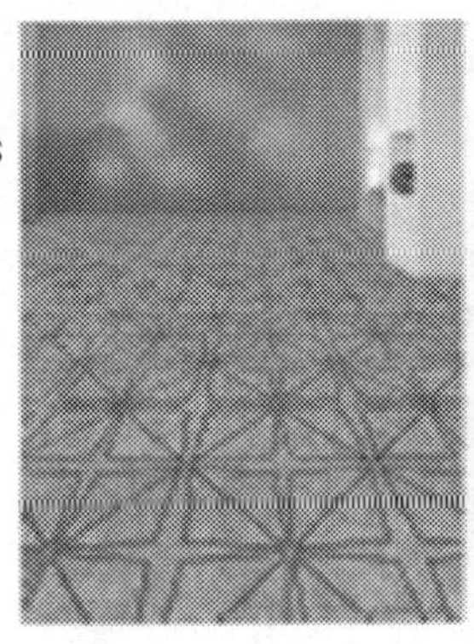

This sensor flooring notices when someone is walking, and recognizes people lying on the floor and triggers an emergency call to the centre.

The basic functions of the SensFloor system combined with building automation and burglar alarm in our show house:

1. Fall recognition
2. Orientation light
3. Activity monitoring
4. Automatic light switch
5. Localization of intruders
6. Water-leak detection

Individual add-on modules for later installation:

- Remote SensFloor maintenance
- Automatic doors
- Energy conservation
- Protection to stop demented residents from running away

While SensFloor offers a fantastic solution the infrastructure requirement is somewhat limiting. The Holy Grail in sensors and wearable tech is being able to integrate both clinical sensor technology with off-the-shelf sensor technology that is HIPPA compliant and available to the doctors, caregivers, families and residents. This data can be used to proactively deal with a crisis before it becomes a crisis. Finally, a company called www.myhealthconnection.tv has developed a comprehensive solution for senior living. My Health Connection can integrate data from a Fitbit or other wearable health tracker to a clinical device and allow anyone needing to see the data to have access. They can also set up alarms to ensure that caregivers and family members can intercede and ultimately avoid a crisis health event. This solution is HIPPA compliant, cost effective, simple to operate and flexible with many add-on modules. The best part is it does not require any infrastructure.

Google Car

Google has decided to change how we drive. Lidar (light radar), GPS and real-time 3D mapping technology from AutoDesk (a software company that is used in architecture and design to create electronic floor plans and 3D modeling of buildings, bridges and roads) have created a self-driving car that should have you rethinking the concept of transportation.

If you have not had the chance to watch the Google Car video on YouTube yet, it's a must-see. Autonomous cars will be one of the largest game-changers in the U.S. in the next 15 years. Currently they are already legal in California and Nevada. While the Google Car is really about resident independence, it requires several technologies that we have discussed or will discuss in the coming chapters to make it all happen. Sensors, artificial intelligence, crowdsourcing, data mining and infinite computing are just a few.

The car uses real-time information that is fed from the Lidar (which sits on top of the car) to create 3D maps that are then tied to GPS with safety preferences built in, similar to several cars on the market that have crash avoidance. The basic premise of the car is that this real-time "map" of the environment records everything from the cars around it and their speed and distance to a dog on the sidewalk to a building to a ball that rolls into the street. In simple terms, this data is then fed into a computer that compares the information to the location the driver has selected and it tells the car what to do … speed up, slow down, break, turn, etc. It's basically like one of those very sophisticated parallel parking options on a luxury car except on steroids.

Peter Diamandis, founder of the XPRIZE, in his top tech tips for 2015 makes **autonomous vehicles #3:** "In 2015, we will see incredible developments in autonomous vehicle technology. Beyond Google, many major car brands are working on autonomous solutions. At CES, Volkswagen will bring the number of car brands on display into double figures for the first time this year. Companies like Mercedes say they will show off a new self-driving concept car that allows its passengers to face each other. BMW plans to show how one of its cars can be set to park itself via a smartwatch app. And, Tesla, of course, has already demonstrated "autopilot" on its Model D."

What does all this mean for senior living? Can you imagine instead of the buses in senior living communities, that you have four, five Google cars and residents can go out as they want, not on a mass trip, and have independence and be safe? I know I want one of these for my teenager who's driving. I have personally witnessed the tragedy of my 19-year-old nephew's passing due to falling asleep at the wheel. Imagine with an autonomous car, texting and sight issues would be a thing of the past. I think the roads would be a lot safer and the data substantiates this. Studies show that you can get up to eight times more cars on the highway safely with autonomous vehicles. Just taking into consideration insurance companies' thoughts, you can see why this is moving along so fast. And that's not to mention, as we grow as a country, the infrastructure would need to have a massive rework unless we can get more cars on the same roads safely. The numbers just work. Vehicle-to-Vehicle Communication Systems, or V2V, will change how we all drive. According to the U.S. National Highway Traffic Safety Administration, they predict that integrating V2V will reduce annual car accidents by 80 percent, with goals for mandatory manufacturer integration planned for 2017.

Now Google knows this and has targeted seniors, people with disabilities, and those that need flexible transportation but could find better things to do with their time than drive.

When Google tapped the YouTube video, some of the responses were priceless and worth hearing. The video starts out with a designer that is explaining to the first test riders outside of Google that they are special. Of special note is that you have two female seniors, a senior married couple, a sight-impaired senior male, and a mother with her small child.

The Google designer explains, "It was a big decision for us to go and start building our purpose-built vehicles. And really they're prototype vehicles. This was a chance for us to explore what it really means to have a self-driving vehicle. But in the small amount of time we've been working on it, we have functional prototypes and that's exciting."

2 Senior Females

Female Speaker: Oh, it's really cool! It's like really kind of a Space Age experience.

Female Speaker: OK. Hurray!

Female Speaker: We're like queens.

Female Speaker: You sit, relax, you don't need to do nothing. It knows when it needs to stop, it knows when it needs to go.

Female Speaker: It actually rides better than my own car.

Female Speaker: Yes.

Married Senior Couple

Male Speaker: What she really liked was that it slowed down before it went around a curve. And then it accelerated in the curve. She's always trying to get me to do it that way.

Female Speaker: That's the way I learned in high school driver's ed.

Sight-Impaired Senior Male

Male Speaker: The human feeling of it is very well-engineered and it is very smooth. There is nothing that makes you feel the least bit threatened. It's impressive! I'm totally in love with this whole concept.

Male Speaker: Our lives are made up of lots and lots of little things, and a lot of those little things for most people have to do with getting from place to place, in order to connect and do things, and be with people, go places that they need to go and do things. And so there is a big part of my life that's missing, and there is a big part of my life that a self-driving vehicle would bring back to me.

Google Team Speaker: This is a first step for us, and it's really exciting to see the progress we've made. The opportunity for people to just move around and not worry about it, it's going to be incredibly empowering and incredibly powerful for people.

Male Speaker: I love this!

The impact on senior living is far-reaching. Imagine seniors staying in their homes longer with a Google Car or similar autonomous vehicle, as they can get to their doctors' appointments, card club, dinner with friends and their grandchild's sporting events safely. Independence is maintained and safety for all involved is increased.

While it will take some time for other states to make it legal, we can anticipate that because California and Nevada have already legalized autonomous vehicles, Arizona would be next ...

Honda has been at the forefront in transportation technology, developing devices such as personal mobility devices using gyroscope technology. What's really neat about these transportation devices is that it's a blend between a Segway and a scooter that seniors use in the U.S. While a Segway could be used by a senior, they would have to stand the entire time; these you sit on, so they're more appropriate for seniors than the Segway. Honda also has another version that's smaller and is light enough to carry with you and just use it when you need it.

The benefit to seniors and senior living is simple: imagine replacing your scooters with these mobility devices – how much less storage would you need at mealtime to store these versus scooters? Or let's say you want to do a trip to the mall. Seniors that would normally use a scooter could take these mobility devices with them due to the size and enjoy getting out more, which would keep them more engaged, making for happier residents.

Core strength would have to be improved in the seniors that would use this device, as slouching and couch-potato posture won't cut it. If you have ever been on a Segway, you know they are incredibly intuitive; they are much easier to learn how to operate than a bike or a car. I anticipate that active independent living communities will be the first to adopt these devices, as well as seniors who are still living at home. Imagine mom's knees give her trouble but she would love to go to the zoo with her grandkids, on a vacation or just a walk with her friends. She feels guilty about slowing them down or the hassle of trying to find a wheelchair when they get to their destination. With mobility devices such as Hondas, all of those worries are gone. Mom can come without any worries and, if need be, fit the device in the overhead of the plane. Freedom and dignity all wrapped up in a nice little bow!

Exoskeletons

While the merging of man and machine has entertained us for decades, one of my favorite TV shows in the '70s was called *The Six Million Dollar Man*. It programmed us to believe that man and machine would happen sooner rather than later. I can still remember hearing the sound they would play when Steve Austin was slow-motion running! While it's taken quite a while to get to a real-life Steve Austin or Bionic Woman, exoskeletons have arrived. Some exoskeletons are even controlled by thought!

Exoskeletons combine quite a few technologies to work, some of which are crowdsourcing, infinite computing and sensors. The important thing to remember throughout this entire book regarding technology is that it rarely stands on its own – it's interconnected and typically requires multiple technologies to be successful, so when thinking of adopting one of these technologies, it will be important to look at what other technologies are connected and how this will impact your execution strategy.

The first pacemaker was implanted in 1958, around the same time cochlear implants were developed. Now folks previously deaf can hear for the first time due to this merge between man and machine.

This topic is so large that at the opening ceremonies of the World Cup Soccer Tournament in Brazil 2014, the first ball of the opening ceremonies was kicked by a paralyzed individual named Juliano Pinto. Walk Again, the group that made it all possible, is an international collaboration with over 100 scientists. What's shocking about this? The exoskeleton is mind-controlled through a helmet that reads the wearer's electrical activity and translates this "intention" into action. In addition, in order to provide feedback to the user, they have created CellulARSkin that provides feedback to the wearer so they know when their foot touches the ground.

While this is groundbreaking, it is the just the beginning in mobility-assistive devices.

The U.S. military has long been a developer of the exoskeleton. Exoskeletons create a unique advantage for a soldier but can also be used to transport wounded soldiers quickly and safely to medical triage units. If you want to have some fun and see what they are really up to, look at the DARPA (The Defense Advanced Research Projects Agency) YouTube videos available. Further developments in these devices use brain control versus muscles or nerves.

U.S. firemen and emergency services have also employed the use of exoskeletons. Imagine for a moment that you are tasked with rescuing a family from a burning building several flights up. You're in excellent physical shape and have trained to deal with all the potential issues. You know that you may need to carry out this family, who may be unconscious. Now let's add equipment that you need to wear to not only be safe but be able to fight the fire: your fire suit (pants, jacket, boots, helmet) your air tank, hoses and extra air. This adds approximately 100-130 lbs. of additional weight on your body that you will have to take up the stairs, and then some of this will still have to be on you when you rescue the family coming down the stairs and out of the building via the fastest route possible.

The human skeleton was not mean to bear excessive weight on it to this degree. Muscles and ligaments are meant to take the pressure off of the skeleton so that it can move freely. "The Hulk" provides extra muscle that works by allowing all the extra weight to be transferred to the skeleton, bypassing the user and transferring it to the ground. Providing the user with Hulk-like strength, it allows the mission of saving the family to be accomplished with less stress and more power and reduces the damage on the hero's body that would normally take place under this kind of wear and tear.

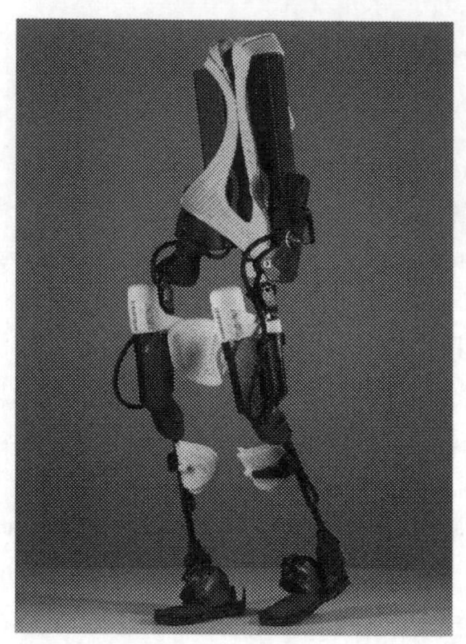

Image Courtesy of 3D Systems

Japan has had exoskeletons for rent for several years now. The intention was for the frail-bodied to be able to maintain their independence yet not be in a wheelchair. By far you will see that the Japanese have pushed harder on robotics than any other country to alleviate issues for their seniors. In comes HAL (Hybrid Assistive Limb), developed at Tsukuba University by engineering professor Yoshiyuki Sankai. The robot suit employs technology that reads your muscle impulses and delivers robotic assistance that can increase strength up to five times. The strength can be dialed up or down according to the needs of the user. HAL is now available to be rented monthly or weekly

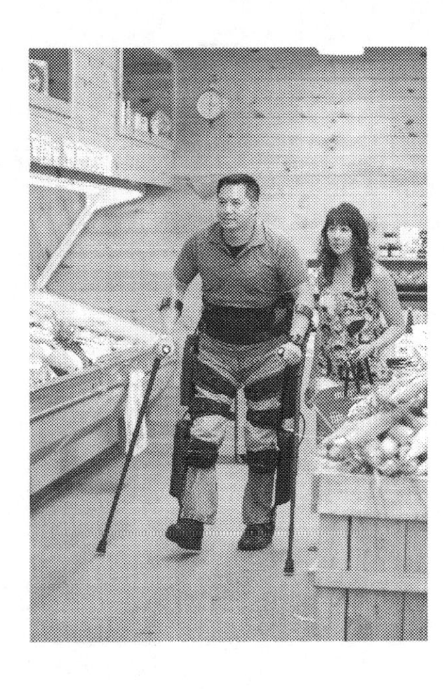

Recently, the FDA approved its first exoskeleton called The ReWalk for public use in the U.S. The implications are earth-shattering to senior living environments. Can you imagine a home without canes, walkers, wheelchairs, and scooters? How would the activity level go up? I can assume damage to walls would go down ... Besides the obvious benefits to physical independence for a senior with mobility issues, can you imagine still being able to walk the dog, go on vacation with the family and grandkids and go up and down stairs easily? As seniors lose abilities and become more sedentary, they become less engaged. The less engaged they are in their world, the less their world engages with them. This downward spiral affects mental health as well as physical. Our minds and bodies were not meant to sit still all day, and when they do, our energy is drained out of us and erodes our value. Everyone wants to be valued, and stopping the downward spiral that is created by being constrained to a wheelchair allows a senior to engage in life, add value and be valued.

The ReWalk not only allows the user to get up from a seated position and walk and then sit back down, but it can help a user get up and down stairs! This is life-changing stuff. In a recent story on the ReWalk with CBS, Dr. Alan J. Kozlowski states that "they are essentially a wearable robot."

Basically, the exoskeleton works together with a computer and motors to help power the knees and hips to see patterns and assist with the movement when body weight is shifted. Dr. Ann Spungen states, "All the terrible things that go wrong in a sedentary lifestyle are magnified with a person with spinal cord injury, so any activity we can provide for them stands to reason to have benefits."

Robert Woo is a 46-year-old who was paralyzed from the waist down when seven tons of steel fell into his construction trailer. He has been able to walk again due to ReWalk, a robotic exoskeleton. Robert Woo stated in an interview with CBS News, "I didn't think I could be useful and I wanted to die. The worst thing is, I couldn't even pull the plug."

Robert Woo put it perfectly. "To be able to stand up next to my wife and give her a hug for the first time, to be able to walk with my children to the park. These are things that we take for granted and I missed."

The cost could be the key: currently the ReWalk costs $70,000. Until the price point hits around $20,000, one could assume it won't be a mass-market item for seniors. But as we saw with scooters, once the reimbursement was figured out, it was not that folks did not want them, it was that they needed to be able to afford them.

Currently, falls could be eliminated with the new stabilization technology that is being employed by the Japanese Robot Asimo or with Segway's gyroscope.

The real benefit for exoskeletons in the near future is for caregivers. An average back injury workers' compensation claim runs around $80,000. With an exoskeleton suite that caregivers can share, this could substantially reduce the risk of injury and still have a return on investment. Not only is this a benefit to the caregiver and your bottom line, but in

> With an exoskeleton suit that caregivers can share, risk could substantially reduced. Caregiver injury would be reduced, resulting in an excellent return on investment. Not only is this a benefit to the caregiver and your bottom line but in resident dignity.

resident dignity. Currently the American system of dealing with a resident that is too heavy utilizes a technique borrowed from longshoremen. Just as containers are hoisted with cranes onto and off of large ships to trains or tractor trailers, we hoist up our residents in Hoyer lifts to transfer them from the bed to a wheelchair or into a tub for bathing. The process is not only undignified but also scary for the resident, especially when they are naked for bathing. Japan sees this as a great opportunity for their caregivers, but also for their factory workers. Being able to lift 75 lbs. without flinching nor risk of injury sounds like a safe bet on tasks that still require too much movement for a robot to complete.

While this technology is available today, it's a safe bet that it will move to the manufacturing sector first, emergency services and then acute healthcare. At this point, the bugs should be worked out and the price point low enough that it will have a major impact on senior living in 5-10 years. Move over Steve Austin, Iron Man has arrived and is no longer a fantasy …

People with peripheral neuropathy (balance, walking and foot-health problems) in America come in around 20 million and most are older adults. WalkJoy believes they have a solution that will help those with peripheral neuropathy to walk normally. By placing a small device below the knee, it helps the person to walk better by stimulating nerves to send a signal to the brain that the person is walking. (Peripheral neuropathy reduces the feeling in your feet, which reduces the information the brain needs to walk properly.) The result: falls are reduced and the person has more confidence to walk, which not only engages them in life more, but gets them moving, which has numerous health benefits.

A recent article on Senior Living Forum titled "*Will Wearable Notification Devices Replace Assisted Living Memory Care?*" Asks the question that families will start asking. Massive investments are being made throughout the world in sensor technology and wearable devices and governments are funding these initiatives. Lok8u is the wearable tech company featured in the article, the device is a watch that sends notifications to the family member or caregiver if the loved one is outside of a zone and at risk of elopement.

For 250 dollars and a 30 dollar a month service agreement families can have peace of mind. The cost is a fraction of the cost of senior living. We all know that it's not an apples to apples comparison but the technology gap is getting smaller.

CHAPTER 4

Induction Looping Technology

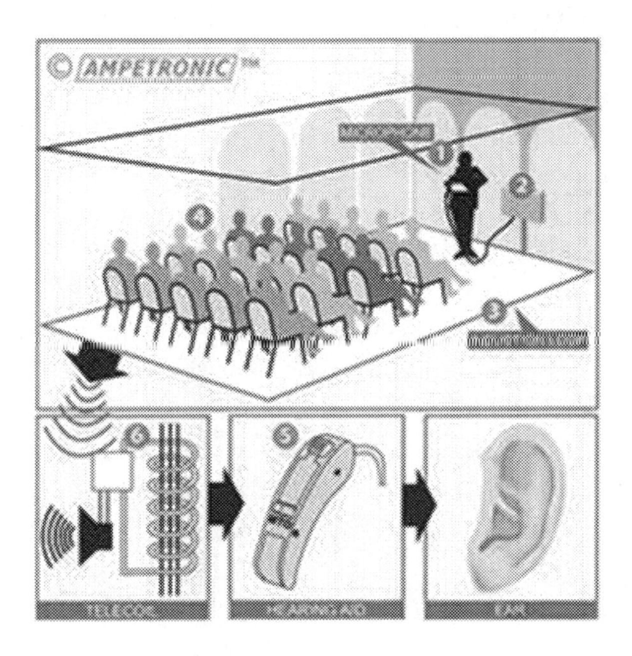

While this chapter is by far the smallest, please don't mistake its size for lack of impact. Induction looping technology has been around for years. The basic premise is that if you are hearing-impaired, even with a hearing aid it is difficult to hear in a large setting such as at church, sporting events, airports, concerts, activities, dances, etc. Basically, it's tough to hear with a hearing aid anywhere a microphone or speaker system is used. Induction looping allows those with a hearing aid to hear directly from the microphone without all the background noise. Louder does not always mean easier to understand.

How it works is a loop of wire is installed in the space that can be small or has large events and uses a speaker system; this can be anywhere from a living room, to a stadium, bingo hall or church. The wire transmits a magnetic signal to a T-coil, which is found in 70 percent of all hearing aids. The hearing aid can then directly hear what is coming out of the microphone, TV or speakers without all the background noise. Europe has been using this technology for years but the U.S. has been slow to adopt. There are some 36 million people with hearing loss in the U.S. and most of them are your residents. Michigan has been the largest initiator in the U.S., having installed induction looping systems in several hundred venues as well as the Detroit airport and Michigan State University's basketball stadium. New York City is adding hearing loops in almost 500 subway information booths and for all future taxi's they will be the equipped with lopping technology and have been branded the "Taxi of Tomorrow." The cost for the technology is relatively small which makes this an excellent affordable technology that can provide maximum positive impact for your residents and staff.

Hearing loss can cause several issues with independence, dignity and safety. Seniors can get paranoid when they can't hear what is being said, or as a coping mechanism they just disengage. They can't hear their favorite shows anymore even when they are blasting, games in the activity room such as bingo or anything with listening becomes a frustration, and church services no longer hold the same value when you can't hear the sermon.

As with most great technology, the looping solution is simple and elegant. You might be thinking why can't Bluetooth solve this issue without the cost of installing a looping system? Bluetooth technology sucks battery power, while the telecoil requires zero. This gives the telecoil a major advantage. Another advantage is that looping can be utilized in both large and small areas while other wireless technologies (such as Bluetooth) can only be used in small areas. Finally, hearing aids all have different proprietary wireless technology depending on the manufacturer. The telecoil is the only universal wireless receiver.

Key Benefits to looping system:

- **Universal** – compatible with any manufacturers hearing aid
- **Affordable** – looping technology is a very low cost technology per space
- **Energy Efficient** – zero drain on the battery
- **Flexible** – can be used in large and small spaces
- **Scalable** – you can start with one space and add on over time
- **Easy** – simple to operate, both for the resident and the staff, reduced listening effort
- **Independent & Convenient** – if looping technology is available the hearing impair person is completely independent of the need to ask for a portable assistive device and then having to plan to pick it up and remember to return it
- **Dignity** – there is no need to ask to be moved closer or get ear phones to hear

Interviews that have occurred after installing the technology will bring you to tears. Seniors speak about how they were able to enjoy the event, hear the sermon for the first time in years or just simply be part of the moment with everyone else, laughing at the joke that was told. In the *Hearing Journal,* Dr. Remensnyder tells of other denominations attending services at St. Mary's of the Annunciation Catholic Church because they have a hearing loop. We all want to engage in life and hearing is one of the most important senses that allow us to do so. Resident engagement is what we all speak about, what if it were as simple as being able to hear the joke, TV show, bingo calls, sermon, or musician? Imagine the impact on your environment and the ripple effect.

David Myers founded HearingLoop.org and has moved the technology to the forefront in Michigan. Other champions of this technology are Dr. Sterkens in from Wisconsin; she is such a believer in getting the word out that she took a year off from her practice to promote the technology. It must be working because the Midwest clearly leads the United States in looping installations. Articles on the benefits of looping technology have now been published in the *New York Times*, *Scientific America*, *National Geographic* and the *Washington Post.*

A recent white paper published in Hearingreview.com October 2014 titled *Consumer Perceptions of the Inductively Looped Venues on the Utility of Their Hearing Devices* stated that "using hearing devices with loop systems **dramatically increases customer satisfaction**." The study confirmed that most consumers complain about the inability to hear speech clearly with their hearing aid in noisy spaces, larger spaces and areas with high ceilings. What was astounding with the studies conducted were the following:

- 99% (approximately) preferred listening with a hearing loop system than without
- 45% of normal hearing people preferred the use of a telecoil in looped venues
- 81% preferred loop with hearing device vs. infrared with neck loop or FM/infrared with headset

Positive statements from the study were as follows: "I didn't have to work so hard to hear the speaker"; "In a loop I don't need to read the captions"; "No more need to read lips"; "**Freeing**"; Feels like I have normal hearing"; "The clarity in a loop is **amazing**"; "I could hear things my wife could not hear"; "**So much better I cried**"; "I would not attend (meetings, church services etc.) if they did not have a loop"; "It is awesome to be able to understand, not just hear. I don't think folks with normal hearing can appreciate just how awesome it is!"

Imagine, with this one simple technology, how your resident satisfaction would improve just by the simple fact that residents could hear what you were saying. TV's could be turned down which would allow staff and residents not to have to raise their voices over them. Engagement in activities would exponentially increase, creating more energy and a higher quality of life for all. The ripple effect will be enormous from families, staff, religious leaders and even entertainment groups.

For more information to find out how you can have this technology implemented into your home, please see this link:
http://www.hearingloop.org/vendors.htm

CHAPTER 5

LED & OLED lighting

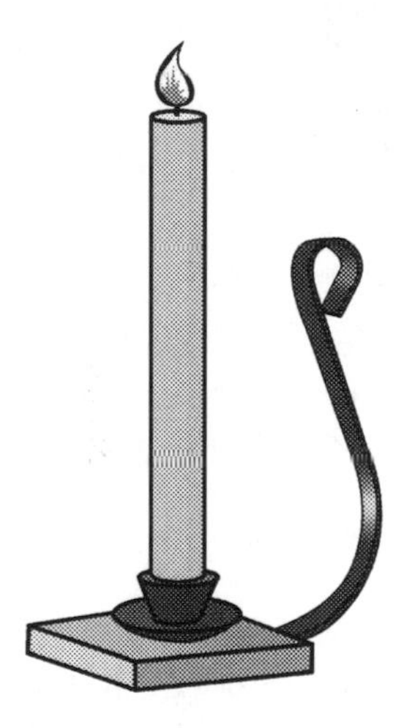

Candles were a main source of lighting during non-daylight hours for centuries. Oil, gas and kerosene lamps created a more stable source of light, but still needed maintenance and could be very dangerous. It was not until 1930's when the incandescent lightbulb became a mainstay in American homes in major cities, but in rural areas only about 10 percent of homes had electricity. By 1939, 25 percent of rural homes had electricity. Even with the low percentage in the rural areas, incandescent lightbulb technology changed forever the way we worked and lived. We were no longer dependent on the sun; our workdays could be longer and factories could add second and third shifts instead of building more factories.

In 1939, my grandfather left Canton, Ohio, to work the World's Fair in New York City. He would tell us stories of all of the movie stars and famous people he guessed at his "guess your weight" stand while at the fair. It was the place to be. It was branded as "Dawn of a New Day" and had an exposition of "the world of tomorrow." TVs were introduced there, as well as nylon, color photography, air-conditioning and fluorescent lamps.

GE held the patent to the first electric bulb. In the next three years, electric bulbs were being produced by many companies and were installed in many manufacturing facilities and open office plans. Not until 1980 did compact fluorescent lights (CFLs) become available to retrofit into the same socket as an incandescent light.

While enough lighting is crucial to seniors (seniors require up to 70 percent more light to see than a 20-year-old) the details are just as important: color temperature, not having bright spots or shadows, reducing the amount of fluorescents due to the eye strain caused by flickering (you may not be able to see it, but the eye still has to adjust), and consistency. Lowering the light levels can make it more difficult to see steps, chair seats and handrails, which causes more stress and fear in the resident. Add in that every individual sees differently with colors and light levels and then throw in for good measure sight issues such as yellowing of the cornea, macular degeneration, nearsightedness and farsightedness, and you can see why developing a well-rounded senior-sensitive lighting package is so critical to quality of life and reducing risk.

CFLs were more energy-efficient but were problematic in wall sconces and ceiling fixtures. Most senior-living budgets are built on a CFL model after the U.S. government created a planned phase-out for incandescent light bulbs.

LED lighting technology has moved faster than expected, with GE and Phillips both moving at the speed of light to develop this technology and bring the price structure down. They have solved the color temperature (how the light makes colors look) issue, which was one of the main reasons that designers shunned the technology; the other was cost. Landscapers have been huge proponents of LED lighting due to its low energy use and long lamp life; color temperature was not a huge concern for them and so on the backs of landscape designers the technology was fine-tuned and the cost became palatable.

We could learn a lot from landscape designers on how to use LEDs to create nighttime theatre that is interesting for the resident and family members without causing issues for sleep, maintenance or energy costs. You may have never noticed it before, but how a building or walkway is lit at night is one of the definers as to how nice we view a property as being. Drive into any exclusive neighborhood or resort and you will see this. Proper landscape lighting creates mood and theatre that is desperately needed in senior living to help bring the outside into the environment.

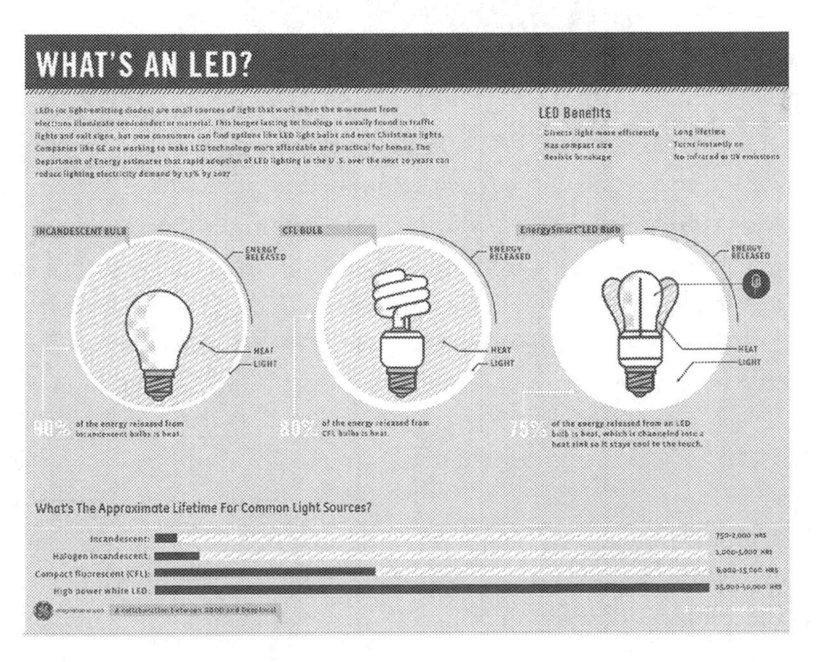

On the interior, senior living has always had residential-style fixtures. These worked well when using incandescent lamps, but as the need rose to retrofit compact fluorescents due to maintenance and energy costs, the end result was a poorly executed Band-Aid approach resulting in compact fluorescent retrofit bulbs sticking out above the shades and covers taking away from the beautifully designed chandeliers and wall scones.

The exciting news on LEDs is that they are not only for new senior living homes but also existing environments. LED lights have been integrated into handrails in showers, door casings, and crown molding, and can change colors to help simulate sunrise, daylight, and sunset to help those with cognitive difficulties.

Creating coffered ceilings with uplighting, or tray ceilings if you prefer, has always been costly due to the structure needed to facilitate the overlapping florescent tubes to avoid dead spaces in the lighting. The crown molding would have to be built out and supported for the lighting. Now with LED lighting, the depth is cut by 50–75 percent, reducing structure, materials and mass. The light can also be programmed to change color, just for fun or to add ambiance.

When LED uplighting is utilized in crown molding in the resident's room, it can be used to simulate the natural light changes from day to night. The lighting is controlled by a computer connection that will sequence sunrise, daytime, sunset and then goes off at night. This simulation is intended to help the resident's circadian rhythm, increasing the melatonin production to reduce sundowners, especially for those with Alzheimer's and dementia. It is suggested to have blackout window treatments to ensure that not too much light can seep into the room at night for the best possible outcomes. They're just getting the data now from an evidence-based design standpoint, but in my experience with designing hotel rooms for United States Air Force pilots, they have to be able to sleep during random hours and blocking out all light is key during this time for their bodies to "feel" like it's time to sleep.

LEDs integrated into flooring, handrails and door casings helps with wayfinding and reduces tripping and falling. These integrated lights can double as nightlights and should be colored amber to avoid keeping a resident awake. EOS LED handrail system shown in the image can not only bear the weight of a resident and be easy to find when your eyes are full of water but also provide a night light for the bathroom.

Imaging a resident room that automatically lights up when you enter because it is tied to motion sensors; the uplighting in the crown molding around the room is the same color temperature as the hour of the day and therefore will change throughout a 24-hour period. The door casing (amber-colored) is lit to help orient the resident at night to minimize the risk of falling, creating a holistic lighting solution that allows technology to assist the resident to be as independent as possible and also reduces operational risk.

The amber LED light embedded linearly into the door casing on the handle side of the door allows the resident to clearly see the door edge and provides enough light to orient themselves to the bathroom door. An LED over the toilet comes on when the resident approaches and gets brighter as they need. Inside the shower and beside the toilet are integrated LED-light handrails that make it easy for the resident to see, especially when their eyes may have water or shampoo in them. For getting out of the shower, a water-resistant LED strip helps to define where the flooring change is to avoid trips and falls. Then, back to the bed, where the sensors take the lights back down and finally off to restful sleep. In the morning the LED cove lighting gently awakens you with the warm colors of a sunrise (with a color temperature of 1850K), changing as the day goes on to a higher-level color temperature between 3000K and 3500K and then back down during sunset. The theory is that, provided this cuing, the body's circadian cycle will be properly set and sundowner's syndrome could be diminished, if not avoided.

The challenge is that most senior living proformas have been built on a cheaper compact fluorescent pricing structure. LEDs are still more expensive than CFLs, but have lower energy costs. Typically, the lighting package on most projects is almost always way lower in dollars than it should be. This is due to the lack of understanding of how many decorative fixtures will be needed as compared to providing basic nursing home lighting and then factoring in the energy codes, which often reduces the amount of fixtures. It's a tough balance when the codes restrict energy use yet seniors need up to 70 percent more light than an individual in their twenties. The perfect solution is LEDs, with the exception on the initial cost. My recommendation would be to double whatever your lighting budget currently is. Great lighting that is not only bright but even and flexible with color and dimming lets you create mood and improve quality of life. Who doesn't like a sunny day vs. a cloudy one?

Being able to reduce energy costs, increase resident independence, improve safety, and create a more residential environment puts LED lighting at the forefront of best technologies for senior living. Adopting this technology is fairly simple and may also have tax credits attached. Just as the fishing net was a simple technology that created greater outcomes than spear fishing, lighting may seem simple but I can assure you the return on investment, both financially and in quality of life are win-win.

OLEDs are organic light-emitting diodes which are different than LEDs (light emitting diodes). The major difference is the organic compounds in OLEDs light up when they get electricity, and LEDs light up because of the movement of the electrons through a semi-conductor. None of this really matters to anyone except for one thing: OLEDs can be extremely thin, bendable and even translucent.

In most categories OLEDs win: black level, lightweight, flexibility, color, response time and viewing angles. LEDs win in size (not for long), brightness and lifespan.

OLEDs are just starting to emerge and will have a larger impact on how we see technology in general. They basically are so small that they can turn a paper-thin layer into a computer or television screen. Imagine turning your entire wall into a TV or your resident charts being able to roll up into the size of a pen and unroll for reading and charting. See the books website for the Samsung video link.

Existing senior living homes can retrofit low ceilings with 1/8" thick OLED panels and raise the ceiling 6" that previously held recessed lighting.

If the ceiling was too low to have recessed lighting and the existing lights were surface-mounted (non-decorative), these panels can be utilized due to the nominal depth creating a more cohesive interior environment.

While senior living has just started to take hold of LEDs and OLEDs, there are numerous simple ways that this incredible technology can be applied. A few examples are: under cabinet lighting, closet lighting, bathroom mirrors with built-in lighting, toilet seats with built-in lighting, backlit menus (I saw these in Germany and they were incredible), back lit room signage, vanity mirrors, cove lighting, ceiling panels and hanger lights.

Besides the obvious energy savings, OLEDs will help to solve one of the largest challenges in senior living, which is proper lighting. Over the past few decades the battle between energy conservation mandated by law and residential-looking fixtures meant that lighting levels were not near what they should be for an 80-year-old. OLEDs are thin enough to fit in wall sconces, cove lighting and retrofit situations, are very energy efficient and deliver excellent light. OLEDs will be the silver bullet for senior-living lighting challenges today and in the future.

© KONICA MINOLTA

Konica Minolta introduced at the Light + Building 2014 Frankfurt trade show their flexible color-tunable OLED lighting panel. These will be mass-produced, which will help to bring the cost down. As these costs are driven down and other players such as Phillips join in, it is anticipated that our imaginations will not be limited by the current confinements of where these light panels go, how they are bent and what can be displayed on them. This will be as great a transformation as the candle was to the lightbulb.

Chapter 6

Crowdsourcing & Gamification

Crowdsourcing seems like a new concept but it's really a quite an old one. Before technology, when it was time to harvest the crops or build a barn, family, friends, neighbors would join forces to help each other out and get large amounts of work completed that would otherwise be impossible. These "crowds" would come together for the greater good or crops would go bad, or help would not come when you needed your barn raised.

Now, with harvest machines and other technologies, we are no longer dependent upon our family, friends, and neighbors to bring in the harvest or raise our barns, but we have new problems, more work than we can handle and micro-expertise that is required to build and create. The solution to these new problems comes in the form of crowdsourcing. With access that has now been created through the Internet we instantly can hire "minions" or micro-experts from anywhere in the world for a fraction of the cost previously available. Angie's List is a great example of crowdsourcing. You can hire folks for tasks and their work is reviewed by the "buyer" so that you know if they do what they say they will do. Not only do you have access to a "crowd" of potential minions, but you become a part of the crowd by providing feedback on the services you bought.

Gamification is the process of applying games to tasks to encourage compliance or fun in doing an activity. You're already familiar with gamification if you or your residents have ever used Wii Fit. Basically, you are playing a game while exercising, which allows you to focus on the fun or the competition vs. the actual workout.

Another example of gamification that you may not be aware of is fantasy football, which is a multibillion-dollar-a-year industry. The NFL noticed that viewers were not tuning in as much as they used to (probably due to the enormous amount of access to entertainment anytime, anywhere). Per Wikipedia: "In 1997, CBS launched the beta version of the first publicly available free fantasy football website. The game immediately became widely popular. Today, it is estimated over 19 million people compete in public and private leagues online nationally."

Engagement is key. How do you get people involved and motivated? Games seem to be the **Holy Grail** of engagement, but not just any games like Bingo and Clue. Games need to be highly interactive, flexible and ever-changing; this is where gamification comes in. People want to not only immerse themselves in a game, but also compete and get rewarded.

At this point you might be wondering what the big deal about crowdsourcing and gamification is. How could this possibly apply to my increasingly complex industry of providing personalized care for higher and higher acuity residents while struggling to get, train and keep staff and make sure that we are compliant with all the codes, regulations and reimbursement models necessary to be successful in senior living?

What if 66 percent of your staff did more of what you wanted them to do through gamification? Or let's take it one step further and your residents were 66 percent more engaged in activities? What if you could reduce your marketing costs by 15 percent by crowdsourcing social media or creating fresh, ever-changing ads? Do you think these results would positively impact the quality of life of your residents and your bottom line? I do! It's very exciting, the world we live in today. Never before have we had this much access to technology to reduce our workload and improve the quality of services we provide.

The reality is there are amazing things happening in the world today, and by the end of this chapter I will have given you enough information to understand how crowdsourcing and gamification can benefit you and your residents and start you thinking about how you will apply these technologies.

Image credited to Nintendo.

Now that we understand the basics of crowdsourcing, let's get into the specifics and review different types, how you access it, and give you some examples. Crowdsourcing takes advantage of connecting specific experts or interested parties to solve challenges in either a competitive or collaborative format.

Crowdsourcing has become huge in the fields of task-related work and specialty creative work such as graphic design, video, social media, writing and coding. Companies such as Fiverr, Task Rabbit, Elance, 99 Designs, Crowdspring and Crowd Flower are just a few and have been growing exponentially year after year.

Personally, we use these companies on a daily basis and save approximately 90 percent on costs we would normally spend if we had not crowdsourced. Don't get me wrong – crowdsourcing requires your effort and engagement, but once you get the hang of it, it's like having a clone. For example, I used to spend anywhere from $2,500 to $4,000 for a logo to be developed. I would get two or three versions and these would be in that particular graphic artist's style. If the artist was not getting what I was trying to accomplish in look or feel I would be out the money and have to research and find another designer and then start the process all over again and spend another $2,500 to $4,000. We had a client that was starting a new senior living company and we were able to use Crowd Spring to create a contest to develop a name for their company and ensure that it was not being used by others. The contest cost $239 and gave us 434 entries to choose from! Then we launched a contest to develop the logo for that same company through 99 Designs for $499. We received 283 entries (different designs). Compare this to two or three versions from one graphic artist and you start to see my point. I had 93 different designers competing for the $499! How it works is you describe in detail what you are looking for and then designers, without being paid, send you their designs, you give feedback and then select a final round of the top entries. All the way through the designs are getting better and better based upon your feedback and sometimes a designer will be "watching" vs. participating and jump in close to the end with an awesome design, similar to how you bid at an auction. It's very much like *The Voice*. At the end, you select the winner and they transfer the logo and all of its rights over to you. One of the largest benefits that I have found, besides the costs, in using this format is that so often what I had in my head as the right direction did not end up being the winner. By seeing so many different options it helps you to really understand what you like and don't like with little risk.

Another example is transcription; we will have a conference call meeting with a client and for $5 on Fiverr it can be transcribed and ready to be put into meeting minutes. I don't know about you, but my team can't transcribe and it costs us a lot more than $5 to type up the audio from a half-hour meeting. My last personal example for you is that using Fiverr, for $5 I was able to have my website analyzed and I found out I had tons of broken links that we were unaware of.

You may wonder how this is all possible. Well, some people do it to get their foot in the door with you, and for others it's a way to make income after their day job. For others it is their actual income. While US$5 will hardly buy you lunch at a fast-food restaurant, US$5 is the daily wage for some countries and $499 for one contest equates to their typical monthly wage. One of the best ways to see if you could utilize crowdsourcing is to have your team list out all their activities in detail and then see which ones you could outsource through crowdsourcing.

Essentially, there are three formats that prevail.

1. Creative Competitions
2. Bid for Task
3. Reward-Based

Creative competitions work great for when you need a logo developed, website designed, article written or software developed. These systems work by signing up to a provider and basically placing a want ad for the work you need completed with as much detail as possible. You can either choose a site where you select how much you are willing to pay guaranteed, and then the creator submits their work and competes for the winning bid, or you allow creators to bid on the work and then you choose the creator that best matches your price point and has the background or portfolio you like the best. The key is feedback: creators need constant feedback to improve, which is easy to do through the secure portal of the crowdsourcing company you have selected.

The upside is that instead of going with your local marketing company and getting the same designer with the same ideas, you could have 60 different designers from all over the world providing you with fresh and sometimes off-base designs. You're able to shorten the design process substantially and reduce your cost. Most crowd creators (77 percent) have a primary job so this is to make extra money, to be creative or just build a reputation.

Utilizing these services can improve your look to the community and keep your events fresh and exciting at a low cost.

On a larger scale, The XPRIZE, created by Peter Diamandis, showed the world that competition creates solutions that were otherwise thought impossible. With NASA downsizing its program and retiring the shuttle, Peter decided to launch a competition similar to the one that Charles Lindbergh and the *Spirit of St. Louis* won by making the first solo non-stop flight from New York to Paris. This contest changed how the world viewed air travel forevermore. Peter created a contest to get a vehicle into space for $100K with three people and that could replicate the trip in two weeks to prove durability for $10 million. Teams from all over the world entered the contest. Twenty-six teams entered and spent over $100 million! Private space travel was born for a fraction of the investment and time that NASA could have done it for. While Peter has created many other XPRIZE contests, the one to watch for in senior living is the QualComm Tricorder XPRIZE. The QualComm Tricorder XPRIZE will soon be awarded and its main goal is to put minor health diagnosis in the hands of the consumer at a higher accuracy rate than that of the current medical profession. Once this takes place, senior living will have to run to catch up to the level of quality and speed in regards to how fast the general public will be able to self-diagnose.

Bid for task works differently and basically can be anything from hiring a handyman, to organizing your bills, to ensuring quality assurance programs.

Let's say you want to shop your homes and see what someone who is not an expert would find, in multiple cities, without hiring a large secret shopper company that would entail a large contract and travel and reimbursable expenses. You could hire secret shoppers on a task site and create a template for exactly what you need them to do where and how you want it reported back to you. You could also do this for your competition at a fraction of the cost. There is one company that will pay you to send inventory data back to them via your camera phone in stores you already shop in. Let's say you're a customer of a national drug store chain, and on your visit you notice they are out of your favorite lipstick; you pull up their app and snap a photo of the shelf. You then notice that the cough drops are in the wrong location and you snap a photo and upload it to their app. Kids are earning $80 a week directly deposited into their accounts for simply being aware in places they already visit and providing valuable asset management data in real time to the store.

Or say you just need to send out handwritten thank-you cards, have an engraving done, custom cookie cutters made, an answering machine message recorded, etc. There are millions of people that have used task crowdsourcing to do these "gigs."

Reward-based crowdsourcing can be more altruistic and/or collaborative, but in nature it's about the work, not a monetary reward.

You could be collaborating with the crowd to identify a DNA sequencing to help find a cure for cancer, or contributing to open-source code development for new software. This category sometimes gets mixed in with gamification and data mining, which we will discuss in greater detail in those chapters. The point is that if you have a problem and want to solve it, don't try to do it by yourself or just with your team; rather, allow others to work with you. In senior living, this could be open problem-solving on developing a standard for senior living that is better for the residents and staff vs. being forced to comply with ADA. ADA was developed for the 20-year-old Vietnam veterans returning from war, not an 85-year-old senior with macular degeneration, arthritis and 40 percent less muscle mass than that of a 20-year-old. Reward-based crowdsourcing would seem to be the perfect fit for collaboration between operators, designers, residents and regulators to get a viable code to follow developed just for the seniors we serve instead of the square peg, round hole system we are currently handcuffed to.

An example of reward-based crowdsourcing is Cellvation (gamified crowdsourcing), a project that was completed in 24 hours to see the possibilities of what kind of impact a reward-based crowdsourcing app could have. Cellvation states that 428 million hours are spent playing online games each day. Their idea was to play a game and save lives in the process.

Cellvation decided to figure out how to use gamification and crowdsourcing to solve malaria and cancer. Every day there are 600,000 cases of malaria. It takes 30 minutes for a clinician to look at a slide. That's 300,000 man-hours spent every day just on diagnosing malaria. Malaria is easily resolved if it's able to be diagnosed properly and quickly. To further complicate the issue, most of the cases are in Africa, where there are not as many clinicians as in developed nations.

The solution: play a game where you identify malaria. With 22 amateur people looking at the slide, they can get 99 percent accuracy and these people are just plain having fun playing the game. After one minute of training, they can go down to 13 people. It's the same thing with cancer.

Prescreening of cancer is said to help prevent the deaths of somewhere between 3 to 35 percent of Americans. Even if they can get prescreened, there's just not enough time or clinicians to do the prescreening. Imagine if we could potentially avoid between 3 to 35 percent of cancer fatalities just by playing a game. There's no money involved, but there are rewards and fun and knowing that you are doing something good. You can hardly speak to someone that has not lost a loved one to cancer ... 584,000 people are estimated to have died of cancer in the U.S. in 2014. Three percent equates to 17,520 lives that could have been saved on the low side and 204,400 lives on the high side! How awesome would this be? Not to mention, the savings in medical costs to the families and system.

Foldit, on the other hand, is not a test but a real game that is helping to save lives. Drugs get tested on computer simulations of human proteins (amino acids) and in a lot of cases the limitations of computers to fold these proteins makes this a better task for humans (score one for the humans!). Per Foldit's website, protein structure is the key. "Protein structure prediction**:** As described above, knowing the structure of a protein is key to understanding how it works and to targeting it with drugs. A small protein can consist of 100 amino acids, while some human proteins can be huge (1000 amino acids). The number of different ways even a small protein can fold is astronomical because there are so many degrees of freedom. Figuring out which of the many, many possible structures is the best one is regarded as one of the hardest problems in biology today and current methods take a lot of money and time, even for computers. Foldit attempts to predict the structure of a protein by taking advantage of humans' puzzle-solving intuitions and having people play competitively to fold the best proteins."

The concept of dealing with disease and drug creation is very similar to a war strategy. The more you have intel on the enemy, know where they are, how many troops they have and what type of flexibility, weapons and weaknesses they many have, the better chance you have at winning the war. Folding proteins helps scientists better understand how proteins function and they can then design strategies or drugs to combat them if they have gone to the dark side and now cause disease.

Per Foldit's site, they explain the relevance to not only HIV/AIDS and cancer but also Alzheimer's, which senior living is all too familiar with:

- **HIV / AIDS:** The HIV virus is made up largely of proteins, and once inside a cell it creates other proteins to help itself reproduce. HIV-1 protease and reverse transcriptase are two proteins made by the HIV virus that help it infect the body and replicate itself. HIV-1 protease cuts the "polyprotein" made by the replicating virus into the functional pieces it needs. Reverse transcriptase converts HIV's genes from RNA into a form its host understands: DNA. Both proteins are critical for the virus to replicate inside the body, and both are targeted by anti-HIV drugs. This is an example of a disease producing proteins that do not occur naturally in the body to help it attack our cells.
- **Cancer:** Cancer is very different from HIV in that it's usually our own proteins to blame, instead of proteins from an outside invader. Cancer arises from the uncontrolled growth of cells in some part of our bodies, such as the lung, breast, or skin. Ordinarily, there are systems of proteins that limit cell growth, but they may be damaged by things like UV rays from the sun or chemicals from cigarette smoke. But other proteins, like p53 tumor suppressor, normally recognize the damage and stop the cell from becoming cancerous — unless they too are damaged. In fact, damage to the gene for p53 occurs in about half of human cancers (together with damage to various other genes).
- **Alzheimer's:** In some ways, Alzheimer's is the disease most directly caused by proteins. A protein called amyloid-beta precursor protein is a normal part of healthy, functioning nerve cells in the

brain. But to do its job, it gets cut into two pieces, leaving behind a little scrap from the middle — amyloid-beta peptide. Many copies of this peptide (short protein segment) can come together to form clumps of protein in the brain. Although many things about Alzheimer's are still not understood, it is thought that these clumps of protein are a major part of the disease.

Most of the major breakthroughs in medicine have come from crowd collaboration of researchers, sharing their breakthroughs and then playing one idea off of the other to create a better solution. Now that the Internet no longer limits researchers to geographic barriers, the rate of collaboration has gone up exponentially.

"You can't teach an old dog new tricks" was a saying I was brought up on and implied that once we become mature, the learning stops. We are cemented to be who we are and our ability to learn becomes very difficult. To further this, we were taught up until 2005 that if something happened to your brain, that part was basically dead and could not be recovered. Senior living lived this firsthand with cognitive issues with their residents and a knowledge that once the cognitive decline started there was little that could be done to recover those faculties.

However, in 2005 a study was released that found that when medical students were studied during a period of studying for their exams, their gray matter actually increased. This then became known as a mainstream concept called neuroplasticity.

The concept, simplified, is that if you exercise your brain, you can keep it healthy just as you would your body, and deter the effects of aging and possibly even reverse them. Several doctors and futurists have written books on this concept and numerous companies are cashing in on "brain games" not only for seniors but for those that fear they will end up like their senile grandparents. The word "senile" is hardly used any longer, but it refers to a person having or showing weaknesses or diseases of old age, especially a loss of mental faculties.

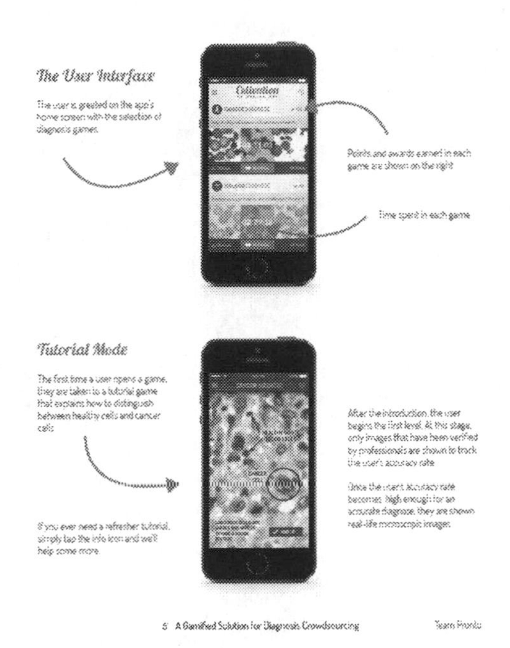

Lumosity is for athletes, it's for kids, it's for seniors, and it's for you and me. Elevate, which I actually play, is free and it's an app. It does the same thing as Lumosity, essentially. Brain Fit Life is software that not only helps your brain to stay fit by playing games, it also provides support and creates a brain report.

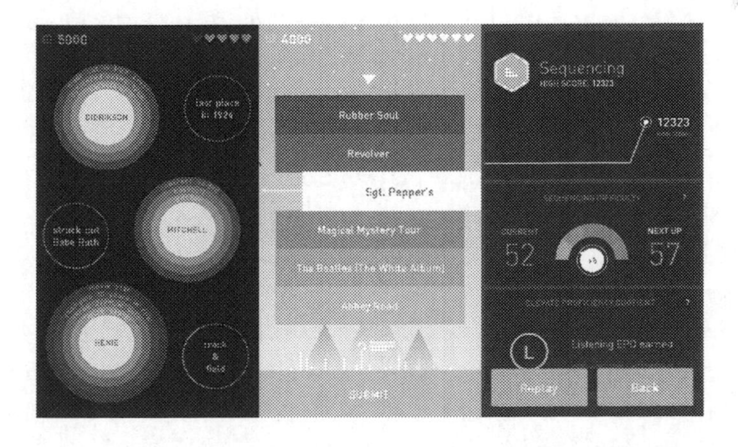

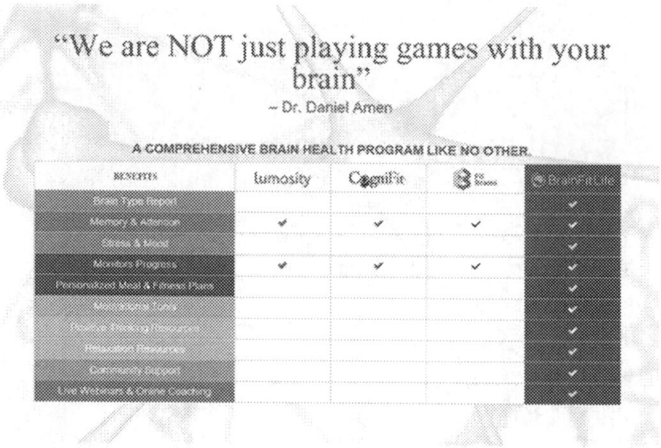

“We are NOT just playing games with your brain”

~ Dr. Daniel Amen

A COMPREHENSIVE BRAIN HEALTH PROGRAM LIKE NO OTHER.

BENEFITS	lumosity	CogniFit	[illegible]	BrainFitLife
Brain Type Report				✓
Memory & Attention	✓	✓	✓	✓
Stress & Mood				✓
Monitors Progress	✓	✓	✓	✓
Personalized Meal & Fitness Plans				✓
[illegible]				✓
Positive Thinking Resources				✓
Relaxation Resources				✓
Community Support				✓
Live Webinars & Online Coaching				✓

Dr. Daniel Amen, one of the leading psychiatrics on brain health, speaks about the fact that psychiatry is the only medical field that guesses to diagnose instead of using the technology we have today in imaging and blood work that any other doctor would use to help them create a proper diagnosis so that they can create the most successful treatment strategy. According to Dr. Mehmet Oz, “Dr. Amen is one of the most gifted minds in medicine.”

He uses the example of Abraham Lincoln being diagnosed as depressed based upon his symptoms of being depressed. The astonishing fact is that this was 173 years ago and the way diagnosis is made today by most doctors is the same as it was then.

While Dr. Amen has been on the forefront of NFL concussion research and debate and helping to heal players from traumatic brain injuries, he is concerned about all brain health and has written books not only on ADD/ADHD but also aging, including *Use Your Brain to Change Your Age: Secrets to Look, Feel, and Think Younger Every Day.*

The relevance to senior living is that Alzheimer’s and dementia in most seniors is diagnosed by symptoms. Rarely is a brain scan completed or labs taken to determine if there is another cause for the memory loss or if the loss could be reversed through supplements and/or medication, TMS, psychological testing, neurofeedback, lifestyle changes, bioidentical hormone replacement therapy, oxygen therapy, hypnotherapy, IV nutrient therapy, prolotherapy or platelet-rich plasma therapy. Not all of these therapies are for memory loss — some are for depression, anxiety, joint aches, etc. Sound like anything your residents have?

Currently, if you're in your 80s and you start to have memory loss, you're written off from really finding out what the issue is and if there's a possibility you could be "fixed" back to a higher level of cognitive function. SPECT scans that look at brain blood flow and activity patterns have been show to change the treatment plan or diagnosis 78.9 percent of the time vs. just asking the patient for their symptoms. This translates to almost 8 of out 10 treatment plans being wrong when not using the technology that we currently have at our disposal.

A study was performed on 8,000 individuals and it was found that older adults have less active brains. Getting your brain back in shape helps you to think clearer and avoid the effects of mental aging. Dr. Amen has developed a system to get not only your brain fit, but a holistic system for diet, understanding your brain type and then brain training that helps to reduce the risk of Alzheimer's disease.

The 2014 Alzheimer's Disease Facts and Figures Report states that "An estimated 5.2 million Americans of all ages have Alzheimer's disease in 2014. This includes an estimated 5 million people age 65 and older. One in nine people age 65 and older (11 percent) has Alzheimer's disease. About one-third of people age 85 and older (32 percent) have Alzheimer's disease. Of those with Alzheimer's disease, the vast majority (82 percent) are age 75 or older."

With these numbers developing a new strategy is critical, as the drugs thus far developed for Alzheimer's disease have made little impact.

Simple programs to engage the brain, proper brain nutrition and utilizing technology to identify issues and aid in early detection will be the key to solving this epidemic.

"Anyone who stops learning is old, whether at twenty or eighty. Anyone who keeps learning stays young. The greatest thing in life is to keep your mind young." —Henry Ford

Volkswagen's Fun Theory is a group that sets out to change behaviors by utilizing fun or gamification. They have completed numerous experiments with throwing trash away, taking the stairs vs. the escalator, and even wearing your seatbelt. If you have flown lately on Delta Airlines, you would have notice that their safety video is actually fun and it helps you to pay attention. While we can't make everything fun, there is a lot more we can do to manipulate behavior to get the desired results we want, all while having fun. If you have children or have ever babysat, you know how to play this game ... from potty training, to making the food on a spoon become an airplane or train with actual sound effects for the mouth to open and the food to go in. Don't get me wrong; I am not implying that we should treat residents and staff as toddlers. What I am saying is that if a process has to be completed, such as charting or taking medications, why can't we make it a little fun, or gamify the process? We can build in rewards to encourage the outcomes we want so that it becomes a win-win, something that is not only fun but challenging and gets us excited. This one technology, if applied properly, could dramatically improve the outcomes in almost every aspect of what you do. It could be as simple as, for every activity that a resident participates in, they receive points or a reward of some sort. We are all competitive at some level, even if it's just against our own goals. Therapists do these micro-goals every day when working with residents; what if it was actually integrated into a system that could reward the resident? Having personally run Wii Bowling competitions, where senior homes came to a central offsite auditorium and competed for prizes, I found seniors to be incredibly competitive, way more than my team had bargained for ☺.

Some of Volkswagen's Fun Theory experiments in how to manipulate behavior for the better and have fun are eye-opening to say they least, and I would highly encourage you to watch the videos with your staff and brainstorm how you may engage the concept.

The Fun Theory site states, "This site is dedicated to the thought that something as simple as fun is the easiest way to change people's behavior for the better. Be it for yourself, for the environment, or for something entirely different, the only thing that matters is that it's change for the better." They encourage participation by offering a fun theory award and testing the top winner's theories.

The most famous of the experiments involves a piano staircase, in which folks are encouraged to take the staircase vs. the escalator out of the subway. They recorded data with cameras of how many individuals took the stairs vs. the escalators and then came in and installed a weight-activated piano on the staircase like the one in the Tom Hanks movie *Big*. Sound would come out as people stepped on the keys of the stairs. Some would try to play a song on the way out of the subway while others tried to hold back a smile. The end result was there was a 66 percent increase in people taking the stairs vs. the escalator!

Seatbelt compliance in children — the idea has been tested in Sweden and works by showing a video to the children in the back seat on putting on their seatbelt. If they don't comply, the video system will not work for them to watch movies.

Another example of the Fun Theory was "The World's Deepest Bin." They took a trash can in a park and rigged a motion sensor sound to it so that when trash was thrown into the bin or can, it made a sound like it never hit the bottom. People would actually go looking for more trash on the ground to throw into the bin just to hear the sound again or to try and figure it out. In the end, 158 lbs. were collected in the trash bin in one day, which was 90 lbs. more than the next-closest bin.

The final Fun Theory example was through a contest. The contest winner postulated that you could reduce speeding by entering those obeying the speed limit into a lottery to win some of money gained by those who were speeding and issued tickets. The concept was simple: get more people to obey the speed limit by giving them a chance to win money from those that broke the law. The results were not surprising: speeding was reduced by 22 percent. Apparently positive reinforcement works better than negative.

Given the results, it seems ridiculous not to employ more of the Fun Theory in our senior living homes, not only for residents but also for staff and families. Who would not want a 66 percent positive increase in participation of any item that we already do such as charting, medication compliance, activities, survey results, etc. Imagine what this would do to positively impact your bottom line, but also the quality of life of your residents.

CHAPTER 7
Robots

Robots range from a Roomba vacuum, introduced in 2002, that can vacuum your floors to those we see in beer manufacturing facilities that have gone from hundreds of employees down to 10. For the purposes of senior living we will be looking at two different types of robots: one is companion robots and the other is specialty robots. As we have done in the other chapters, I will give you the latest technology out there and then we will dive deeper into what is and will be applicable for senior living. As mentioned in previous chapters, this technology, like the others, is heavily dependent on multiple other technologies. Robotics utilizes data mining and artificial intelligence, sensors and infinite computing.

Per Wikipedia: "In 2012, the growth of robotic lawn mower sales was 15 times that of the traditional styles.[3] With the emergence of smart phones some robotic mowers have integrated features within custom apps to adjust settings or scheduled mowing times and frequency, as well as manually control the mower with a digital joystick."

Jibo is touted as the "world's first family robot" which, as this book prints, is being released as a table-top robot that can be your phone, talk to you, help you find items on the Internet and remind you of your schedule, take family photos, act as a companion as it recognizes you and greets you, can read you a story or the newspaper, etc. We are moving into a world where the Rosie the Robot from the cartoon *The Jetsons* is becoming reality. Siri has lessened our apprehension of talking to non-live beings to the point that we are comfortable with them materializing into an actual thing that we coexist with in our homes and work.

In order for you to understand how far these technologies have been developed, I am going to take you to a place that may seem farfetched, but I can assure you it's real. Just as at the 1939 World's Fair in New York they had to show a TV with a clear screen so that folks could see the inner workings of the TV because they felt they were being tricked and had trouble understanding that the TV was actually projecting the president's speech live, just like it was in radio except with pictures, I am going to pull back the curtain and show you what the world has been working on the last several decades. It is my hope that you will recognize robots are here to stay and will be integrated into our lives in the very near future, not unlike the sci-fi movies that we've seen.

Darpa is the government's research and development (R&D) arm that helps to further along technologies that have real-world applications in the U.S. military. The robots that I will be discussing through Darpa are public knowledge and the implications are far-reaching. They basically host contests that are very specific to what they want to accomplish, and several companies will develop technologies to try to win the contract (did you hear the see the word "contests"?). That's correct: they use crowdsourcing to accomplish what would either take too long to develop on their own or would be too expensive, through contests similar to the XPRIZE that we discussed in a previous chapter.

While I am sure they have many top-secret technologies and contests, the ones that are public are incredible. A few these awesome inventions follow.

- Cheetah – is a 4 legged robot that can run 29 miles per hour
- BigDog –by Boston Dynamics (Purchased by Google in 2013) is a robot pack mule
- Petman and the later version Atlas – Humanoid Robot that can perform search and rescue.

- Sandflea – think of the remote controlled cars you give at Christmas, now attach a camera and have it jump 30 feet straight up into the air.
- RHex – similar to SandFlea, Rhex is a six legged remote controlled device the size of a toy car that is used to gain intel safely

These robots are created to save lives by accessing hostage situations, gaining intelligence, conducting search and rescue, and performing mission critical operations. While anyone could argue that the government could create an army of robots to use against us, we also understand that the term Robot = mechanical or virtual artificial agent per Wikipedia and basically means that we are utilizing technology to perform tasks that are simple, repetitive, dangerous or require immense strength. Imagine that your son or daughter (who is serving in the military) is able deploy a robot that is operated by them from the safety of a base in the United States. They are able to perform a military action instead of being in "theatre," as it is called when you are in the warzone. This is the intention of the technology: to keep our servicemen and women safe while they are protecting our freedom.

Manufacturing has been using robotics for decades. Now, it is utilizing not only robots but also exoskeletons to reduce the impact on the human body when a process still requires the intelligence or nuance of a human to perform the task at the highest level.

Robots are an excellent workforce as they don't get sick and call off work, have "issues," ask for a raise, get tired, need holidays off, get bored by repetitive work, form a union, create sexual harassment issues … well, you get the point — the list can go on and on. It makes sense to convert what can be converted to robotics if the task and the cost for the investment make sense.

A factory in China, Foxconn, the world's largest manufacturing company, was having difficulty with workers who were complaining, becoming ill and even committing suicide. They were going to strike for better benefits, etc., even after management addressed many issues. Foxconn management installed robots and let the workers know that by that time next year the entire factory would be robotic if they could not work out the issues. Suddenly the issues stopped and the factory was back to normal. Robots help to control outcomes. Foxconn has decided, though, that placating workers is more difficult than it's worth and will be installing one million robots over the next three years. Robots cannot yet (thank gosh) understand subtleties. We are entering an era in robotics equal to that of the previous industrial revolution, where the local tailor feared that the factory would take away his business of making suits, and he was right to a certain extent. It's no different than the travel agent and the Internet. Things change and improve, but there will always be work to do. Maybe not the same work, but someone has to design the robots, maintain them and program them. Workers will have to up their skills to become masters of this new technology (machinery). A tailor would no more go back to hand-sewing a suit than an architect would go back to hand-drafting his design documents for your building. The benefits of technology are far too great not to embrace them.

One of the most exciting developments in robotic technology for factories is the flexibility of programming. Imagine that robotics currently is like the first-generation computers, where you had to code what you wanted it to do just to create a simple contact list of all of your business associates (if you're under 40, this analogy will be lost on you). Now with the recent advances, you can literally teach your robotic friend via placing his hand where you want it to go and the series of actions you would like him to do step by step, and then hit save and voilà! The robot Baxter is programmed and then will fill in the blanks. The implications of this are astounding as now line workers can modify quickly the robot to be flexible to change line runs and improve cost, quality and performance without having to invest huge amounts of time to program or being held to the rigidity of having one robot that is built for "X" but now you need it to do "Y."

As promised, I am going to get to the really far-out sci-fi robots that are, well … let's put it nicely: creepy.

Japan and Denmark have two of the top scientists in the world developing Geminoids. What are Geminoids? Essentially they are a duplicate copy of a human in robotic form but not a clone. The premise is that the Geminoid can interact for you even if you are not able to be in the actual space. They are a tele-operated robot of an actual person. Similar to the Beam where your Skype type image is able to move around on wheels, a Geminoid actually looks like the person and speaks. If you have ever been to Disney and witnessed Abraham Lincoln speaking and moving through the magic of Audio-Animatronics technology, it's essentially the same thing but on steroids.

Japan and Denmark have entered into collaboration and recently hosted in Denmark the JST CREST / Patient@home SOSU Future Lab Workshop on portable androids and its applications in March of 2015.

Per the website "JST **CREST** (Core Research of Evolutional Science & Technology) research project"Studies on Cellphone-type Teleoperated Androids Transmitting Human Presence" is going to finish its five-year research on teleoperated androids. The aim of this study is to develop cellphone-type teleoperated androids that enable us to transmit our presence, anywhere and anytime. A user transmits his/her presence to a remote place and the partner in distant location can talk to him/her while feeling as if they are facing each other. Such new information media have been designed to harmonize humans with information environment beyond existing personal computers and cellphones. This study explores its potential as new human-harmonized communication media through social experiments in various countries such as Japan and Denmark."

Demarks Patient@home project is testing these robots for use in their healthcare.

Of special interest during the conference was a session on seniors. The description of the session is as follows "**Portable androids for aged citizens** -The study on the portable androids explores their applications in field experiments in which people interact with one another by their mediation. With the focus on the social aspects of androids that may facilitate human communications, the Telenoid has been applied to dementia care in order to observe the elderly's natural reactions to the robot and to develop service models and new communication media. This session reports the results from Japan and Denmark, and discusses the androids' potential for aged citizens of different nationalities."

Six professors / researchers from Japan, Denmark, Germany and US presented their views and results utilizing Telenoids with seniors (aged citizens). A session was even dedicated to Dementia – "Humanoid robots in dementia-care: Investigating if Telenoid alleviates symptoms of dementia"

Hiroshi Ishiguro from Japan has developed not only a Geminoid that looks just like him but has also developed a female copy of a woman.

Henrik Scharfe from Denmark has also produced a Geminoid that looks just like him and states that he has become attached to having him around. His Geminoid project made him one of *Time* magazine's 100 most influential people in 2012.

It gets even creepier as they have found a way to get a microchip to merge with DNA and function as one, learning and mimicking. It's rumored that someone is looking to take his DNA and merge it with his Geminoid to create a robot clone of himself. Does anyone else think this was a plot in a recent movie? Others are working nonstop to have artificial intelligence (A.I.) in a Geminoid form and have successfully developed an A.I. that can think and have conversations. Phillip – is an AI robot that can build a model of who he is talking to with facial and speech recognition and learns every day. He seems to have a sense of humor which is one of the most difficult things for an AI to accomplish. While the YouTube video is interesting it's also unnerving. Experts remind us that AI is all around us, just think of your spam filter. While we can understand simple AI and Abraham Lincoln at Disney it's hard to make the leap to having a copy of yourself in the closet waiting to be booted up to talk to your kids while you're away on business.

Why, why, why would anyone want to do this, and what could be the potential benefits? As I have been told, the reasons are to help reduce pain and suffering in humanity and, get this ... they specifically mention seniors as one of the target markets!

The first way they intend on using Geminoids to reduce pain and suffering in humanity is to provide them as an alternative to a live person for sexual use. Yes, I am serious. Now, stay with me. While you may not be aware, the sex trade is enormous in the world and sex trafficking is a real thing. The movie *Taken* with Liam Neeson exposed this world to mainstream America, but I can assure you, after sitting on the board of Trade Justice Mission, this is a real crisis. Girls and boys are being stolen, sold and tricked into sex trafficking, and once they are in, they almost never get out. In Columbus, Ohio, you can get a girl as young as six years old. As horrific and disgusting as this is (a portion of the book sales will go towards helping to stop sex trafficking and to research for Alzheimer's dementia), it's real. The thought is that they could substantially reduce the need for trafficking if these Geminoids were available without legal implications. If they would save even a handful of kids, I am all for it.

Now on to a less controversial use of Geminoids. Say your husband dies after 50 years of marriage. You miss him greatly, and also miss how he would help you around the house. You could order up a version of your husband (could be younger or the same age as you) and he would be your companion, help you carry out your daily routine and have the strength of Superman and never leave the seat up! Sounds really bizarre, right? I am not so sure; couples that have lived together for many years almost feel as though they have died when a spouse does. They are yin and yang, and when the other is gone it's as if a piece of them is missing. You can see this taking hold in countries that are more accepting of technology and robots than America, but the point is, it could be accepted. It's similar to looking at pictures, watching a movie or listening to a voice recording of a loved one that has passed. While you know they are no longer with you, the sounds and visual provide great comfort.

Geminoids have been tested in hospitals to see how patients would react to them and it seems that they are, in fact, creepy. People were uncomfortable with how lifelike they were, yet not human. It's a very similar dilemma that the computer-generated movie industry has experienced. We are completely OK with watching *Star Wars* and the creations that are imagined, as well as *Toy Story* and the characters that are not real, but as soon as we take this to a human form it becomes a little too real but still apparent it's not real, which creates the "creepy" factor. Computer-animated movies such as Disney's *Mars Needs Moms* were unsuccessful because they overstepped the line. This line is constantly moving, but at the moment the Geminoids are disturbing at best.

Now that I have set the stage and you understand how far-reaching robots have become in our world, I am going to review robotic technology that is being utilized now in senior living globally and how it's not so much about the package but the benefit. Robots come in all different shapes and sizes, from furry little seals, to big plastic bears, to iPads on wheels.

Hector is a robot that is available in Europe. He is used in both smart homes for dementia and Alzheimer's residents and as a companion robot for those that can still be on their own but need a little help. He reminds residents of their appointments, their classes, when to eat and what to do. He can also Skype with other people, get your doctor and play games with you. Hector moves the environment in a very similar fashion to the Beam device and utilizes yes-like movements to engage you, similar to Jibo, but has a facial look more similar to humans. Jibo looks more like a minion.

The lines are blurring. Robots interact with you and not only respond, but have features that blur the line between human and robot. They are no longer being perceived as a computer software program but as a real friend. When we connect on a deeper level, amazing things can happen. We come alive. Emotions bring us out of our shell and we can laugh, get competitive, appreciate and enjoy our quality of life on a much deeper level.

Don't believe me? Let's use a real-world example, I live in Columbus, Ohio, the home of The Ohio State University where a college football team battles for the top spot every year. Now, if you're not an Ohio State fan or happen to even be from the state that begins with an "M" up north, please humor me. Hundreds of thousands of fans from all over the world watch their season and believe they have an impact in the Buckeyes' success. In 2014, the Buckeyes seemed like they would have a great year when their quarterback Braxton Miller was hurt, then one of the players committed suicide, then their second-string quarterback J.T. Barrett broke his ankle in the 4th quarter of the game against the team up north.. Their third-string quarterback Cardale Jones had to finish the game. They made it to the final four and their third-string quarterback was playing his first game as a Buckeye on a national stage against Nick Saban's Alabama Crimson Tide and what do you know, they won! During this game, my family sat in the living room and watched the game and screamed, cheered and felt the adrenaline rush as we pulled out a win. We were all alive! Fast-forward to the National Championship game and my family was mandated to sit in the same seats, wear the same outfits, etc. for fear that we would lose. And that loss would not be due to the team or the coaching, but our family sitting at home watching the game. After The Ohio State Buckeyes won the National Championship in 2015, I heard numerous stories from Buckeyes fans all over the country about how they participated in the winning of the National Championship. Now, common sense says that my position on the sofa, nor my father's clothing, nor me cheering at the top of my lungs at the TV, had anything to do with the outcome, but we want to believe it does. When a resident sees the robot respond with voice and eye movement, the lines get blurred, not unlike when you're cheering the team on and then they score a touchdown. This positive reinforcement helps to reduce fear and produces the type of brain chemicals that create connection and pleasure. The trick is that we pick up when we think the connection is contrived or repetitive and it becomes less real. Our very real investment in watching TV is more than just time; we invest our emotions in this technology and long for our favorite shows to come on. Just as it was difficult for people to imagine this connection when the first television was introduced at the World's Fair in 1939, it is difficult for us to understand the impact this new technology will have.

Paro is a Japanese robot that is actually being used now in the U.S. to work with Alzheimer's and Dementia residents. It costs about $6,500. It's a small baby seal that looks like a stuffed animal and has big deer-like eyes and white fur. It's about the size of a baby doll and interacts with the residents both with motion of the body, motion of the eyes blinking and sound. It's extremely non-threatening and can be petted. Studies with Paro have found that residents that would not previously engage or talk, come out of their shell. One of the major benefits that Paro has over a dog or cat is that if the resident is a bit too harsh, Paro doesn't mind. You don't have to worry about Paro biting anyone or having to take him outside. While there has been large success with life-skill stations (such as a nursery station) in memory care to get a resident to come out of his/her shell, typically a caregiver still has to initiate the activity. Paro takes this on and reacts to the residents and engages them with positive distraction. The only issue at the moment is the cost. It's difficult to get approval for a $6,500 robotic stuffed animal unless the entire home has bought off on the benefit.

A more affordable option may be on the horizon. Milo is a robot with a child's face that is under two feet tall that has been used in therapy with great success with autistic children. Milo is now being tested with Alzheimer's and Dementia residents and shows great promise. Milo's benefits are not only that his cost is substantially less than Paro but that Milo can carry on a conversation, ask a resident to exercise with them and even tell stories. Milo does not get irritated when the same question is asked over and over again and speaks a bit slower than a person so he is easier to understand. He does have a wild hairdo but time will tell if this becomes an issue for older adults. An even more engaging but less touchable technology is Gerijoy's dog and cat avatars. Essentially, through your tablet (iPad) or computer the senior develops a relationship with their chosen cartoon dog or cat avatar on the screen that has a live person on the other end (women in the Philippines). Through previously loaded information such as photos likes and dislikes or needs, the avatar can carry on a real-time conversation! Because the avatar's voice is computerized the senior can chat anytime they want and feel they are connected to their dog or cat (avatar) vs. being confused by an accent or different voice. The dog or cat can show them photos, check up on them if they haven't engaged in a while, and remind them to drink water or just talk.

Gerijoy is just one of many current technologies that is aligned with Aging2.0. If you are not aware of Aging2.0 I would recommend you look into the organization and their conferences. They are the first to merge the thought of aging and technology.

While the U.S. is certainly making strides in robotics and technology for seniors, the Japanese have been obsessed with robots and this has culminated not only in the creation of the Geminoids that we discussed earlier, but in Honda's Asimo. Asimo is by far the coolest, non-creepy robot around. He can dance, climb stairs and now even run! Honda has painstakingly made Asimo's look and stature in such a way that we do not fear him. He is smaller than a human and has a similar body structure, but looks like he is in a space suit ready to land on the moon. His movements are so fluid that when he dances, you think he has to be human. He has rules built into him that make him step aside if he's crossing paths with a human. One of his most amazing feats of late is that he can run five and a half miles per hour. What's the big deal, you ask? Both of his feet have to leave the ground at the same time and then he has to rebalance and to land. This is something that a robot could not do previously. It's amazing how many things you and I have the ability to do that we think are simple, but to a computer they are earth-shattering.

At Honda's headquarters, you can see multiple Asimos in the company cafeteria waiting on staff, taking their orders for coffee, making it and bringing it back to them. Asimo has facial recognition that allows him to not only greet you by name and know your favorites, but also remember who got what when he brings the orders back to the table. If you have seen the movie *I, Robot*, Asimo is very close to that reality.

The Japanese have unique reasons for developing their robotic technology not only for manufacturing but for senior living. They have the largest population of centurions, with a growing population of women entering the workforce and not marrying. Traditionally, the daughter would take care of the elder. The Chinese are starting to experience the same issues as the result of the one child policy, which creates the perfect storm for not having enough young people to take care of the aged. Asian cultures have a high respect for their senior population and are scrambling to figure out how to meet their needs with these new challenges.

A simpler version of Hector that is basically Skype on wheels is The Beam. It was created to answer the call of all the companies that cut travel when the economy experienced a dip, yet still needed to interact with multiple teams and be there in person. The issue with Skype or GoToMeeting is that you are stagnant, only seeing what others want you to see and not being fully immersed in the situation to be able to add value.

A beam-like device was featured on the TV show *The Big Bang Theory* when Sheldon, the resident genius, decided it was too risky to expose himself to other humans. The wow factor of The Beam is that you can drive it and control where you are and who or what you are looking at.

Imaging the following scenarios and if The Beam would be of value to your team, doctors, residents or their families.

Scenario 1: Family and Residents – What if your families could sign out the Beam and go for a walk with their mother, saying hi to her friends and even going to dinner or therapy with her? All while they are sitting at their laptop miles away. Gone would be the days of having to arrange a time to Skype (which is a pain for the staff and families), the staff getting mom to the Skype room and then just talking versus engaging.

Scenario 2: Doctor – What if your doctors could sign out the Beam and do rounds with the residents. They could watch them walk to see how they are doing and then see another resident without traveling to the site. While this cannot replace the touch factor that helps doctors to diagnose, the U.S.

Military has been using telemedicine from everything from diagnosis to surgery.

Scenario 3: Quality Assurance and Emergency Response – What if your regional team has to cover multiple properties regarding everything from operations, maintenance, construction & renovation to marketing and you have a Beam at each property. A pipe breaks during a snow storm and the companies' expert is four hours away, he would just power up the Beam and can drive it to the location and problem solve while looking at the issue all from a remote location. Or your VP of Marketing wants to see what the building looks like at a moment's notice to understand why the numbers are the way they are vs. the team being able to prep for their visit.

There are too many scenarios to name where having an onsite, drivable, interactive device could not only save dollars and time but also reputation. The most unique factor is that you drive the Beam, moving it at your will. When this happens the lines get blurred between robot and human and people within minutes treat the Beam as the person who is operating it.

Beams currently cost is about $15,000 for a commercial model and are used in many businesses where travel is a cost or time issue. They have just started to ship a $1,500 model for the home for seniors. We have been testing the commercial model for over a year and can't wait to get our home model in a month to see the value.

Drones have been on the news constantly, typically flying over sporting events, providing unique views that were otherwise previously impossible unless you owned a blimp. The have filmed fireworks and live events. Farmers are using them to see their land without having to physically examine inch by inch, and real estate is using them to get more up-to-date images than Google Earth can provide. They're being used for sports, crime control, firefighting, and delivering items through Amazon and even pizza by Dominos.

The impact on seniors is immense. One company out of San Francisco is delivering prescriptions through a drone system. Pharmacy items for $1 on top of the script? Because of this, a senior can stay in their home and have their medication delivered to them. They may drive, but when the weather is poor or let's say they have not invested the money to have a Google car yet, it can be very convenient. One of the largest pains in the neck is getting your prescription filled when you don't feel well or you don't have access to transportation. This is just one example, but I can see Meals on Wheels going to drone delivery and having kids volunteer to pilot them on their lunch hour.

Robots have been around in some form or another for decades. I have a personal relationship with my ATM: it knows my name, can tell me details about myself, and if it's in the right mood, will even reward me if I ask nicely for money. Manufacturing has replaced many people's jobs with robotics, especially when the task is repetitive. Robots are intended to move us up Maslow's hierarchy of needs, allowing us to focus on quality of life instead of survival.

In senior living, the robots we need to focus on are companions, caregivers, mobility devices, and safety and connection tools.

The Japanese, as always, are at the forefront of robotics and are the ones to watch in how to best utilize this technology for senior living.

CHAPTER 8

Data Mining and Artificial Intelligence

One technology that surrounds you every day but you may not be aware of is data mining. Facebook and Google use it every day. When you get on the Internet, whether you're aware of this or not, they are targeting you. They are sending you things that you've already looked at so it's specifically targeted toward your likes.

It's meant to learn your behaviors and patterns and reduce the amount of digital noise so that you can actually live a life, purchase and buy from them. That's their ultimate goal. If you have recently looked up how to quit smoking, this gets "mined" and then you see ads scrolling on the side of Facebook for e-cigarettes, patches and any other means to help you quit smoking. It's really quite brilliant, but also a little unnerving.

In senior living, although we have tons of data on residents, we do nothing with it, zilch, zero. This is not necessarily a HIPAA issue. It's an "us" issue. We are just starting to get all of our medical records converted over to an electronic format. Some of the benefits of data mining in senior living will not just be in the realm of improving health. If you have ever stayed at a Four Seasons Hotel, they are known for data mining as a way to improve customer service. They don't use Google that I am aware of in the hotel, but use earpieces to communicate with each other. It goes something like this: when a guest arrives, they have an idea when they will be coming. They use clues like a spy would to figure out who they are and, if need be, the bellman will ask their name. The bellman then radios the front desk and lets them know Mr. Smith is coming in. The guest is greeted with a warm welcome to the Four Seasons. "Mr. Smith, we're so glad you're staying with us." This process keeps happening through the whole stay. How awesome would a potential new resident feel if we used this same kind of data mining in senior living? As I have mentioned previously, not all technology involves fancy space-age computers; some technology that we already have access to just needs to be better utilized.

The best way to explain the practical applications of data mining and artificial intelligence is to meet IBM's Watson. Watson is a supercomputer that can learn. One of the most difficult things for a computer to do is to recognize natural language or human speech and be able to respond appropriately to it. It has to understand the context and the double meanings that can be confusing even for us humans. This is one of the main reasons why it was such a breakthrough when Watson, in 2011, beat the top Jeopardy winners on national TV.

In 2013, WellPoint, IBM, Watson and Memorial Sloan-Kettering Cancer Center started working together on the first A.I. application for treatment decisions in lung cancer.

WellPoint (Anthem) has utilized the supercomputer to increase patient outcomes and reduce costs. While this may seem scary and out of a sci-fi movie (the merger of the Evil Health Insurance company with the Evil Artificial Intelligence Computer), it's actually quite the opposite. What's more is that because of the constant improvement to Watson's knowledge base by participating doctors and nurses, *"Watson's successful diagnosis for lung cancer is 90%, compared to 50% for human doctors."* Now let me ask you a simple question: If you or a loved one was thought to have lung cancer, who would you rather get your diagnosis from? A doctor, who would diagnosis as well as I could with the toss of a coin, or Watson, that is 90 percent accurate? Earlier, better diagnosis not only improves outcomes but saves money. Now imagine this type of technology in senior living, especially memory care, added in with sensor technology to improve medication compliance and increased feedback for medical data.

To further the case for AI and Data Mining, one of the largest issues for healthcare is the immense amount of data that is being constantly updated in the world of healthcare. Healthcare professionals are expected to read, analyze and apply this information in the form of correct diagnosis and treatment for their patients. Watson, however, can not only process one million books in seconds but can also take this information and apply it to help physicians with complex patient diagnosis.

> *"30% of $2.3 trillion dollars spent on healthcare in the US annually is wasted" per WellPoint (The Company is now named Anthem).*

Part of the reason Watson is able to happen is infinite computing. Infinite computing used to just be for the governments. It used to be for the big corporations. Now, you can rent versus own. That means you can have access to faster, cheaper and more accessible computing.

Say you want to create a program or somebody out there has a great idea? Now that individual can have access to the largest fastest computer main frames in the world and only pay for what they need. This is similar to movie stars renting a five million dollar necklace to wear to the Oscars and look fabulous. Things are going to just continue to go faster and faster because now anyone has access to the mainframes. In the near future, we will see an explosion of solutions for providers due to the combination of a global network of software developers and access to infinite computing power. While it will be difficult to sort through the opportunities, imagine reading the want ads in your local newspaper compared to looking at Craigslist online. Tools will be developed to help you find the solutions and the providers you need.

CHAPTER 9

3D Printing

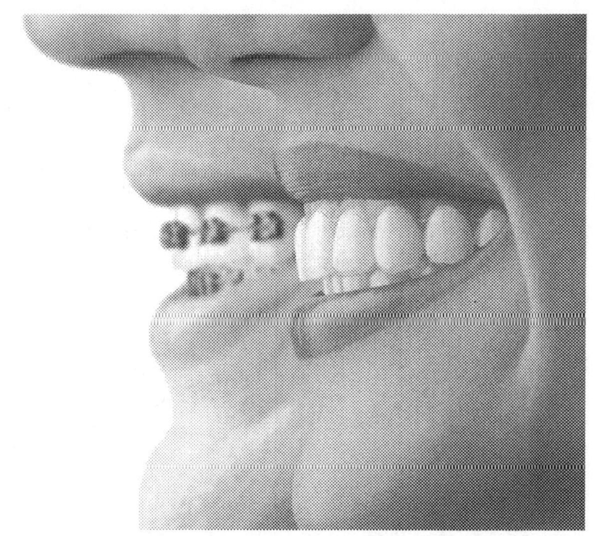

Image provided courtesy of Align Technology Inc.

Did you know that 3D printing has been around for quite some time? You may have seen it and not even noticed it. In 1999, Align Technology changed the world of orthodontics with Invisalign and the technology of 3D printing. This opened up an entire market of adults who wanted braces but were embarrassed of the look. The clear orthodontic trays are manufactured by using 3D printed models representing the desired position your teeth will move. After a 3D scan is made of the wearer's teeth, a computer simulation creates a series of clear orthodontic appliances to move teeth over time.

While you may have been aware that Invisalign clear braces are custom manufactured using 3D printing, you'll be shocked to know companies are now printing organs, human tissue, leather, food, hearing aids, dentures, Barbie clothing, cars, drum kits, candy and metal. One of the most interesting aspects is that they have started printing multiple materials together.

Another enormous implication is prototyping. It used to be that in order to take an idea to reality you would have to design it and hope it was perfect and then have a mold created or a die cast. This could cost tens of thousands of dollars. If you needed to tweak it, you would have to wait until you made your money on the first production run and then make another mold. As you can see, this kept a lot of great inventions from being able to compete with the big boys because the startup costs were large and risky. This was similar to computing power that we discussed in the previous chapter. Now, you can have an idea, create a computer model for it and print it out, and then tweak it hundreds of times before you have ever reached the cost of making a mold.

Image courtesy of 3D Systems

3D Systems can create a geometric shape within a geometric shape within a geometric shape…well, you get the point. Not only the complexity, but that nothing has to be punched out or assembled is amazing!

Dentures can also be 3D printed. I am sure this will be hard to imagine, but humor me. Your resident loses his dentures and instead of having to go to the dentist and go through the entire process with not only the family but the dentist and the wait, you just put in a flash drive and it prints them out a new set!

adidas is a trade mark of the adidas Group used with permission.

Adidas (shown above) has 3D printed athletic shoes, and so do Nike and Reebok. Reebok and Under Armour are hitting the ground running in combining sensors, or what they are terming "wearables," with 3D printing. Some companies are printing shoes and shipping them to you after you custom-design them on a site, while others like United Nude are printing them right in the store.

Why is this relevant to senior living? First, the big boys are in the game and here to stay and see the value. Second, with the knee and hip issues that seniors have, look to having a 3D printing orthotic machine in your home to help residents walk better, which will result in less falls ... Starting to make sense now?

I would imagine that changing the country store or gift shop in your home out to a store that supports a resident's ability to be independent and have a higher quality of life will be bigger than the ice cream parlor was 20 years ago.

When a resident loses his hearing aid, it's difficult for everyone. Instead of having to go to the doctor and go through the entire process with not only the family but the doctor, you could just put the resident's flash drive into the 3D printer and voila! A new hearing aid.

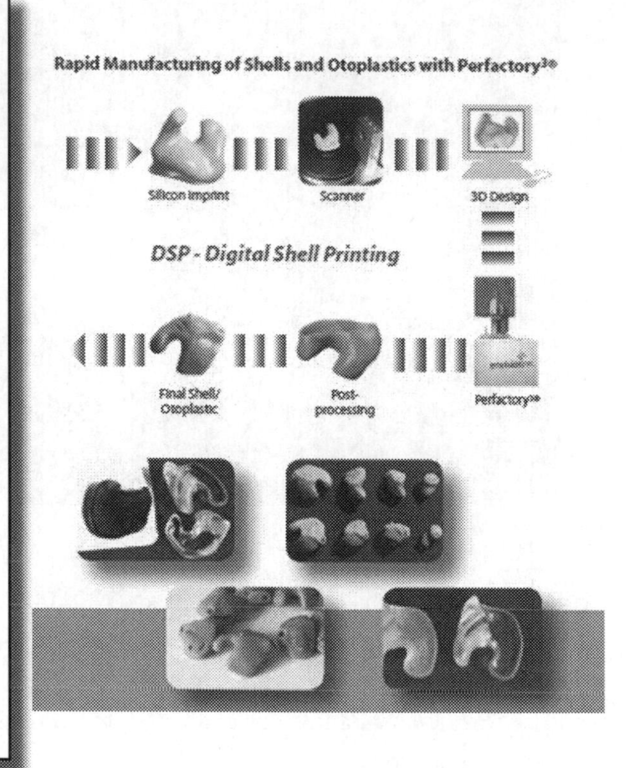

What do 3D printing, Michelin Star chefs and senior living have in common?

Twenty percent of senior living residents are said to have dysphagia or difficulty swallowing. This percentage rises with Alzheimer's and Dementia residents.

From my own personal experience, when I first started working in senior living I was shocked to see that when a resident had difficulty swallowing, the standard method of operation was to take what the meal was for the day and throw it into a blender and then let the resident drink their hamburger. To me it was much worse than baby food, as baby food usually sticks to veggies and fruits. From a resident dignity standpoint, it seemed like we could be doing much better.

Great food options are finally starting to trickle into independent and assisted living due to tracking of resident satisfaction back to food quality, options and temperature. Memory care home-style cooking is becoming more popular, which is wonderful, but this has not addressed the dysphagia issue.

When I asked, what do 3D printing, Michelin Star chefs and senior living have in common, you may not have thought food. There are a couple of 3D food-printing companies making a huge impact in senior living in Europe.

An example of a gelatinous food prepared for the elderly by Biozoon.

Biozoon out of Germany is not only focused on creating foods that look good but are easy to swallow. They have lines that create molecular gastronomy and fancy cocktails. Molecular gastronomy is basically "food science"; it's a modern form of cooking that you would see at a very fancy restaurant or on a TV show like *Iron Chef.* If you're still unfamiliar with what I am referring to, it's things like oak moss vapor, liquid pea sphere, transparent raviolis, etc. This innovative cuisine has birthed new appliances to merge 3D printing and food into everyday cooking. It's really not that amazing when you think about it. If you have ever watched the Discovery's Channel TV show *How It's Made*, you have seen all kinds of food being 3D printed, from Cheetos to the beloved Easter Peeps.

Molecular gastronomy is basically "food science"; it's a modern form of cooking that you would see at a very fancy restaurant or on a TV show like *Iron Chef.* This image is of a liquid pea sphere offered at a high-end restaurant.

The ability to deconstruct food and then reconstruct it has been mastered with these machines, allowing chefs to reimagine what food should look and feel like. We should be thanking all the Michelin Star chefs for teaching senior living that you can make food that tastes excellent but is also easy to swallow and still looks familiar to the resident.

Another company out of Spain called Natural Machines is beta-testing a 3D food printing machine called the Foodini that they want to be in everyone's home. They have created Foodini to be able to handle high-volume restaurants (professional kitchens) as well as a single user.

I was able to interview Lynette Kucsma, the CMO and co-founder of Natural Machines. The following is our conversation, and it should answer the majority of your questions regarding this new technology and how 3D food printing will be applicable to senior living.

Lisa: Can you explain how your product is or is not a home-based product?

Lynette: Yes. We are not just a home-based product. We are targeting kitchen users so it's all types of kitchen users, from professional kitchens to home kitchen users. With the professional kitchens we work with anybody from catering to restaurants to Michelin Star chefs, to even hospitals that are interested and elderly care homes. We work with a lot of Michelin-starred chefs; they are obviously interested in the presentation aspects of 3-D printing, that it actually creates things that really aren't possible by hand.

That's the market we are going after first with Foodini because based on our market research we've found that the household market is interested in a feature where their food can come out of the machine cooked as well, and our first product line does not do this. We are working on that device, but it's going to take us a little bit more time to actually develop it. That's going to be the second device that we release.

Lisa: I am very curious about production time. Is it a lot slower than a normal chef or sous chefs doing the same work in a kitchen?

Lynette: No, with 3D food printing it has to be faster than typical plastic or traditional printing because no one is going to wait hours for dinner, right? We have to make it quite fast. It really depends on what you are printing. It's the material, the food you are using; the quantity, the shape and what quantities you have. You can do things like chocolate sculptures that are a couple of inches high that might take a half-hour, or if you are doing things like crackers or individual ravioli, that can take under a minute to do it.

It really depends on what you are printing. We've always built it with speed in mind as well.

Lisa: How has your team trialed Foodini with different foods to understand the time involved for different amounts of servings? I'm trying to understand how it works with the professional kitchen user that would be in senior living.

Lynette: Yes, there's a few ways I can approach it this question. Basically the way Foodini works is there's room for up to five food capsules to put in the machine at one time. The size of the machine is about the same as a larger microwave, I would say. The capsules will be automatically exchanged as the machine needs them so you don't have to sit there and change out the ingredients. Now again, in terms of speed it really depends on what you are talking about.

One of the nice things that we have with Foodini is that it actually ships with empty stainless-steel food capsules so you are not forced to buy pre-sealed food capsules from us or anybody else.

Lynette: You can put in whatever food you want, whatever combination to accommodate for taste or for medical needs. It's all very fresh food. The texture of the food doesn't have to be all baby food formats, if you will. I'll use an example of a tomato sauce. It can't be too watery, otherwise it's just going to jerk through the capsule system, but it cannot be overly chunky because then it's going to clog up the system and it won't print. But you could have pieces of basil or seasoning at that spine.

The way we envision it is 3D food printing is fairly new for this format. 3D printing has been around since the '80s, but 3D food printing for our countertop model is fairly new. We envision that the range of textures that you can print will, of course, increase as we develop the technology.

Lisa: Regarding the stainless steel capsules, is there any kind of dummy option or strainer kind of thing that says hey, it has to be this consistency, or is by trial and error? How will people learn how thick and how thin to make it?

Lynette: You can learn easily, because we are using a capsule system. We are not using a syringe system, which is what some 3D printers use. One of the reasons for that is to increase speed. If you are doing a syringe system, it's printing out very small amounts at one time. Our stainless steel capsules actually have nozzles that twist on and off, and that have different sizes to them. If you are printing bread dough, for example, you can use a wider nozzle hole than you would if you were printing a tomato sauce as another example. The sizes that we created are 1.5 ml and 4 ml, but we plan on expanding out that range as well.

Again, you'll know if something is way too watery by it dripping through the system, but there are light things you can test with. The way we describe it is kind of like a food processor. When you first get a food processor, you might actually play with the recipes that come with it to understand the machine, but once you understand the machine and the functionality, then you kind of throw away the recipe booklet and start playing with your own recipes. That's what we find people are doing with Foodini as well.

Lisa: What material do you print on? Is it a parchment paper or something similar that goes into the printer so that you are able to pull the food out once it's printed on and then bake it?

Lynette: You can. With this particular model of Foodini we available, it does not cook. It does have a glass tray that you can print on and that glass is like Pyrex so you can stick it in the oven if you want to, but we also tend to print on baking paper so you can slide it right off the glass dish and put it onto your baking tray.

Lisa: Are any of the items that are shown on your website fried, or is everything baked?

Lynette: Well, you do fry some of the food. We can take it from the machine and fry it. I think some of the burgers we fried, and the raviolis we decided to bake them, but we could have fried them if we wanted to.

Lisa: I was looking at the hash browns with sweet potatoes and apple fritters.

Lynette: Those we actually just decided to bake, just because I'm a big fan of baking versus frying. That's just my own personal preference, but you can fry them if you want to.

Lisa: In regards to all the wonderful shapes you show printed on the site, is there a computer software that allows you pick shapes, or how does it know which shape you are supposed to do?

Lynette: Foodini is connected to the Internet. On the front of the device is a touchscreen that allows you to manipulate the machine, which is connected to the cloud. You can actually access those websites from your laptop or your tablet. We created a software called Foodini Creator which allows you to easily import or create your own shapes that you want to print. For example, if you have a logo of your company and you want to print it, you can just grab it from your website and put it into our software and you'll be able to print it.

I don't know if you've ever seen standard 3-D printing software but it's quite complex and intimidating. At the end of the day, Foodini is a kitchen appliance so we needed to make it easy, so we actually had to create our own software to be able to have people do it very easily.

Lisa: When do you think that you are going to release your first Foodini?

Lynette: The first production runs are scheduled to go in the first quarter of 2015 and that's what we are calling a beta, early access release production run. It's to make sure the manufacturing lines are running properly, that they are high-quality devices coming off of the line. We are going to limit the numbers to it and then we will mass market towards the end of 2015. We can't give a hard date because it really is determined on how the machines are coming off the manufacturing line in the beta run.

Lisa: I'm sure everyone will be curious about pricing — what range will Foodini be at?

Lynette: We are looking at about €1,000, which is around US$1,300 right now. It's about the price of a high-end food processor or a high-end coffee maker and it's on the lower end of the price scale for a 3D printer. Actually, the professional market, once they have gotten their hands on Foodini and they understand it, they tell us the price is too low. So we might raise it slightly.

Lisa: Will it be available in both European electric and North American?

Lynette: Yes. It's similar to your laptop, so you can bring that to any country and just have a plug adaptor on it when you are traveling.

Lisa: Is it easy to clean? I know this is always a large concern for senior living.

Lynette: Absolutely; nobody wants a kitchen appliance that is a hassle to clean. We actually built Foodini from day one to be a kitchen appliance, so that means we are building with kitchen-approved materials. We are going through all the UL certifications and similar certifications worldwide so it's a proper kitchen appliance. The actual walls of Foodini are very easy to wipe down. Think of something like your microwave, where it's a very easy type of wall to wipe down. The capsules are stainless steel and they unscrew with the nozzle so you can easily wash it by hand or you can put it in a dishwasher since it's kitchen-grade stainless steel. The glass plate that comes with Foodini is strong glass, so you can throw it in the dishwasher or wash it by hand as well. We are making it really easy to clean.

Lisa: What is the height limit? If you are printing, obviously you have to work within the constraints, but let's say I am making raviolis and then you have five tubes — could you have five raviolis printing at once?

Lynette: With the ravioli example, you would make your own dough and your own filling outside of the machine as if you were making it by hand because our proposition to consumers with 3D food printing is to get people away from all the packaged and processed foods that are in our supermarkets. It's not designed to be the fastest way to get food. The fastest way to get food is to buy something frozen or premade and throw it in the oven.

But it is designed to get people back in their kitchens and making food with fresh ingredients; it speeds up the process of assembly. Our proposition is not to say that every single item you should eat in the future should be 3D printed, just as everything you eat now does not come out of an oven. But those foods that, if you were to make by hand, would require food shaping or forming or layers like the ravioli or making pretzels or making breadsticks, that's where you use 3D food printers with fresh ingredients.

Lisa: So the 3D printer allows you to use fresh ingredients and create complex food with less production time, correct?

Lynette: With the ravioli example, you would actually make the dough and the filling, which is not that complex. The hard part of making homemade ravioli and the time-consuming part is actually assembling that ravioli, rolling out the dough to a thin layer, adding your filling, adding another layer of dough and then cutting it to size. With Foodini you would put your dough in one of the capsules and put your filling in another capsule and then Foodini would print the individual ravioli for you.

It will print the bottom layer first, the bottom layer of dough. Then it will switch the capsules and grab the filling and print all the fillings. Then it will switch it back to the dough and print the top layer of dough for you. You are printing individual raviolis. The speed that we can do that, if you are doing a basic ravioli, basic square shape, it's about a minute and 30 seconds to two minutes per ravioli, with maybe about 30 seconds for exchange.

Again, it's not meant to be super, super speedy, but it's not slow either; you are talking about a minute or two per ravioli, which is not that bad compared to making it by hand.

Lisa: With the size of the interior, how many raviolis could be printed at one time?

Lynette: It's really depending on the size of your plate. The size of the plate is slightly bigger than a microwave plate and it's how much you can fill in the plate.

Lynette: You can print closer together. The height restriction wouldn't come into play with the flat food, but if you are doing a chocolate sculpture or a mashed potato sculpture, then you would have a height restriction of around seven inches.

Lisa: Do you see Foodini as having engagement or entertainment value?

Lynette: Both. Let's say you want to print a spinach quiche in the shape of Batman. You can find a Batman logo from the Internet, pull it down and it makes the shape.

I have two kids. I have a four-year-old and a six-year-old, and it's one of my ways to engage them in cooking because they know it's kind of like playing a game at the end where you are choosing your shapes and printing and then you have different shapes. And as you know, we use our eyes as much as our mouth in enjoying food. This is especially true with kids — especially when you put food in a different shape and suddenly they love it, whereas when it's in a standard shape they might not even touch it.

Lisa: Does the food taste good?

Lynette: Absolutely. That's the thing with Foodini: you are putting in your own fresh, real ingredients, so it's your food. It's not like you are buying presold capsules from us and people aren't sure of the taste. It's whatever you put in. Now, it's not a magic machine, so you cannot put in bad-tasting ingredients and expect great ingredients coming out of the other side. That we haven't quite accomplished yet.

Lisa: It might look pretty, but taste bad if you don't know how to make the interior of ravioli. I get it.

Lynette: Exactly, that's right, but we've had several camera crews come in and take food out to the streets without us to do taste-testing and the only negative responses we get are when people are told it's 3D printed food. They may have the mental hurdle of actually trying the food, but every single person who's tried the food has loved it and then they get over their fear of 3D printing.

Lisa: Can you actually see into it, like a microwave, when it's printing?

Lynette: Yes, you can see it. The top half of the machine is pretty much closed. That's where the electronics are, so you don't necessarily see all the moving bits, and that also protects the machine. It's a high-quality machine but when it's actually printing you can see what it's doing, and as part of the touchscreen interface that we have on the machine we have what we are calling elevator buttons, if you will, that will allow for adjustments. If you see that something is printing a bit too spotty where there's lines in between prints, you can slow down the speed a little bit or you can lower the nozzle to the dish. The reason why we do that is because when you are working with dough, for example, the dough you make today might be slightly different than the dough you make tomorrow because you are using a different flour or a different water. Even though we automate a lot of the settings of Foodini, we do allow for those minor adjustments that, when you are seeing things print, you can actually adjust what it's doing. It's also fun to watch!

Lisa: I can imagine it could be entertaining for kids or seniors and adults the first couple of times they use it, but then after when it starts becoming just a part of what you are doing, you set it and then you walk away and then half an hour later you have your stuff and it's just like anything else.

Lynette: Exactly. We even think in the future you are going to start seeing a mash-up of kitchen appliances. As we are developing a 3D food printer that also cooks, you are getting a mash-up of oven functionality and there's no reason why you can't also mash up a refrigeration technology as well. In theory, in the future you can actually preload your 3D food printer. It'll go into a cooling setting so it keeps your food cool and then when you are on your way home you can use your iPhone or whatever we are using in those days and tell it, on my way home, to start printing dinner and then dinner is printed when you walk in through the door.

Lisa: I see Foodini as an appliance where you could both use it for production purposes but you could also use it for engagement activities.

Lynette: Yes, exactly. We've been getting interest where there's people that just want to open up 3D printing restaurants because they can have 3D printers for show in the middle of the restaurants. One of the things to keep in mind with 3D printing, is what the Michelin star chefs get: whatever dish comes out of the 3D printer doesn't have to be the final dish. You can part 3D print something, part complete it by hand.

Lynette: The 3D print bake can actually do the complicated or designer elements and then you can add food by hand, which also negates the texture issue of a fairly limited amount of textures that you can print right now. That's how you get around that whole issue with different textures on your plate and what have you.

Image courtesy of 3D Systems

As you can see, 3D printing for dentures, hearing aids, clothing and now food will revolutionize how everything is made. We are entering another industrial revolution, but it will be where we can print whatever we want, whenever we want. If you wear glasses, imagine printing out new frames and popping in your lenses if you want a new look or you just broke the frames. Bill Gates imagined that every home would have a computer 37 years ago, and most of us would have said he was nuts. Then imagine a printer in your home. Remember when there were stores that developed photos? Now you either upload them to the cloud to be printed, or print them yourself at home.

Image provided courtesy of Jim Rodda, Zheng Labs.

Everything is changing exponentially. They say if you want to see if a technology will be adopted quickly into the mass market, look at your kids and teens. If it were not for teens, cell phones would have taken much longer to hit critical mass. Right now, little girls are printing their own clothing and accessories for their Barbie dolls with their 3D printers.

You can find files to download for free or purchase on many sites such as Pinterest. This image of the Barbie outfits sells for $30 and that's just for a file to download. Your little girl will still need the 3D printer and printing plastic. Think of a Shrinky Dink or Easy Bake Oven on steroids.

Chapter 10

GE 2025 Future of Appliances

I had the honor of interviewing General Electric's chief designer for appliances, Paul Haney, regarding the vision they have laid out for how technology will impact the home through appliance design.

In GE's Home 2025 Appliances of the Future, in their vision they created personas explaining the needs and then the technology that could help those needs for each persona. While each technology has implications for senior living, Multi-tasking Maria and Reluctant Roberto are GE's senior personas. Maria is 61 years old has raised her family, and has her 84-year-old father living with her. The vision is that the technology can ease Maria's burden and Roberto's guilt of having his daughter take care of him.

While interviewing Paul regarding how quickly these technologies will be available, it was clear that market demand, which will reduce cost, will be the key. The technology is not farfetched. Most of it exists today in various forms. Tying the various technologies together into a seamless system that is cost-effective will be the key to mass adoption.

For instance, GE already has an oven that is a four-in-one unit that uses Advantium technology. Sounds like something out of *Star Trek*, but it's been around since 1999.

According to GE, "Advantium technology harnesses the power of light. The outside of the food is cooked like a conventional oven, with radiant heat produced by halogen bulbs above and below the food. This halogen-produced heat receives a boost of microwave energy. The result? Foods brown and cook evenly and fast, while retaining their natural moisture."

Advantium combines a speed cook oven, convection oven, sensor microwave oven and warming proofing oven all in one. Imagine cooking food in anywhere from half to three-quarters of the time it would normally take with a convection oven vs. the issues involved with microwave cooking. The issue as to why this technology is not in everyone's home is a simple one: price. As with any new technology, research and development gets paid for by the first adopters who are either wealthy or just have to have the latest and greatest and are willing to pay the price. The issue is, this is not a new smart phone that gets changed out every two years, and with new house sales declining, it seems that this technology is destined to stay in the homes of the rich and famous.

While I am sure this technology will start to become more mainstream, let's focus on the future and exciting new possibilities that GE believes will shape how we live.

Coming back to Maria and Roberto, GE imagines how food is grown, delivered and prepared, medicine is delivered and monitored and how we can better connect though everything from our smart phones to our furniture.

Items not identified with the Maria and Roberto personas but other personas that GE created address additional technologies that will have huge value to baby boomers.

They are:

- 3D food printer
- Induction units
- Smart faucet
- Laundry and clothing storage

Paul states that GE has not only worked with ADA guidelines but also Universal Design for their aging-in-place vision. They have engaged in the use of empathetic tools such as sight. Paul explains, "We have a lot of small, little graphics. How do we approach that? One is arthritis, obviously the being able to turn the knobs, open the doors, just simple things like that. The other is reaching up very high or having to get on a step stool. If we design the kitchen from a holistic standpoint, how do you bring things down more into say the nose, top of the head to the waist areas? Picking up heavier loads, that's another issue. We are looking the big picture. Obviously we are not going to be able to solve every problem. We are going to tick off as many as we can."

These are GE's more near-term concerns vs. the 2025 product. They found through their ergonomic studies that seniors can't bend over and get in the back of a dishwasher, or get in the back of a laundry machine. They are trying to solve how they can help seniors achieve their goals around cooking and cleaning to stay independent longer.

One of the largest concerns for a parent staying in their home is them cooking and forgetting to turn the stove off. GE has this issue front and center on their radar. Paul states, "We have a couple of ideas around if you think of a sensor over a stove to where it'll actually have some sort of smoke detector in it that will, if it starts smoking or the heat builds up too high, somebody just leaves it on, automatically shut off the system. That's one idea. Another one is real light ... we have a couple of stoves with light bars over the handle, and those are out now to where they can notify people when things are done. The other is sound. It is obviously a big issue if you lose your hearing; you don't want a siren going off. That's a little bit of a tricky one. What we are doing is exploring, as this generation of baby boomers will move into the last part of their lives, and some of them are really tech-savvy and some of them are still not."

Paul does not feel it's realistic to think that we can just push all of these notifications to their smart phones, as some just won't be that tech-savvy at the point they need help.

One of the other factors that GE is taking seriously is the need to downsize: making appliances the right size for the situation vs. a one-size-fits-all methodology. The challenge is that smaller used to be perceived as lower quality. Think of an apartment-sized refrigerator, 22 inches wide, vs. a 48-inch-wide GE Monogram side-by-side refrigerator. GE believes that while baby boomers will downsize to smaller spaces, they will not give up on quality. It seems we are finally hitting the issues that space-tight countries such as Japan and Europe have been dealing with for decades. How do we not give up on quality and services, yet make our spaces more flexible and functional?

One of GE's micro-kitchen concepts – dubbed the monoblock – is an integrated unit with cooking, dishwashing and refrigeration in a single standalone enclosure that would become a seamless part of the cabinetry.

GE's press release for FirstBuilt states, "'Boomers will have a huge impact on smaller living and it is GE's bet that they won't want to lose any of the luxury or convenience they've had in their lives,' said Lenzi. 'Whether they need a micro kitchen for their downsized dwelling, vacation home, refurbished man cave or boat … Boomers have always wanted the best.'"

In developing what would be the ideal smart kitchen, GE utilized the concept of crowdsourcing that was detailed out in a previous chapter. They basically built a micro-factory called FirstBuild and then engaged a co-creation social media community to help them design the first prototypes.

"Through FirstBuild and its global online community, GE Appliances is able to create, design, build and sell new innovations for your home faster than ever before," says Venkat Venkatakrishnan, director of R&D for GE Appliances and mentor for FirstBuild. "We launched a **micro-kitchen challenge** in May, and everyone from enthusiasts to experts can join FirstBuild.com to contribute their ideas to make the concepts a reality."

Who better than your clients to help you design and develop the product they will eventually buy? As my mentor Dan Sullivan of Strategic Coach says, "Always test your ideas out on check-writers." It seems that GE has done just that.

Concepts range from putting nice cabinetry doors on the front of washers and ventless dryers so that they can be in open view, eliminating the need for a laundry room or the awkwardness of having a stackable in a closet that is sandwiched behind cheap bi-fold doors, to a "monoblock," which is an entire kitchen in a single standalone unit that looks like cabinets.

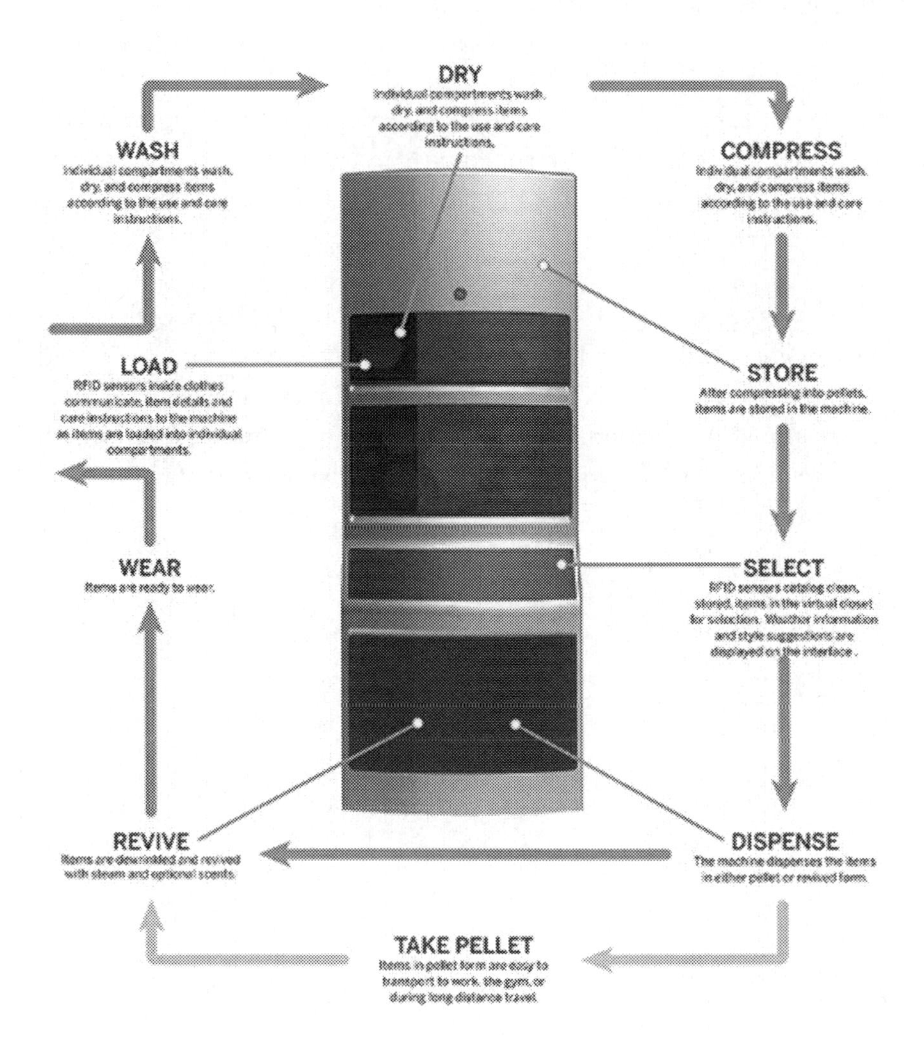
DRY
Individual compartments wash, dry, and compress items according to the use and care instructions.
WASH
Individual compartments wash, dry, and compress items according to the use and care instructions.
COMPRESS
Individual compartments wash, dry, and compress items according to the use and care instructions.
LOAD
RFID sensors inside clothes communicate, item details and care instructions to the machine as items are loaded into individual compartments.
STORE
After compressing into pellets, items are stored in the machine.
WEAR
Items are ready to wear.
SELECT
RFID sensors catalog clean, stored, items in the virtual closet for selection. Weather information and style suggestions are displayed on the interface .
REVIVE
Items are dewrinkled and revived with steam and optional scents.
DISPENSE
The machine dispenses the items in either pellet or revived form.
TAKE PELLET
Items in pellet form are easy to transport to work, the gym, or during long distance travel.

One of the more applicable future technologies GE has for senior living involves laundry. Imagine washing, drying and storing clothes in a single unit.

While GE's concept of an all-in-one washer and dryer has been around for years, the storage component of clothing due to fabric being compressed into a pellet makes this the perfect fit for a senior.

I have used a Korean all-in-one washer and dryer for over 10 years at a lake house. It has one opening and you place the clothes in and then it washes and then dries them, all in one space. It seems to take quite a bit more time per load because the activities are combined. It works great for beach towels and to set and not having to worry about switching the load so it won't get mildewed. While the all-in-one washer and dryer would be perfect for a senior doing one load a day at the most, I can't see this working for a family with kids as you are not able to multi-task and handle large quantities of clothing.

GE's vision is that of that this type of laundry application would take up less room and add more convenience. According to Paul Hanley, "It came from the idea that, if you've ever been to a sporting event where they shoot those T-shirts out at people." This is only possible due to new technology in wrinkle-free fabrics. The implications for senior living are immense …

Imagine for a moment that every resident's seasonal clothing would fit into a shoebox! One of the largest complaints, besides food, in senior living is lack of storage space. This could reduce the size of a closet by 50 to 75 percent, which would free space up for the room or allow residents not to have to choose what clothes they will have to get rid of, which can be a very dramatic event.

Let's develop this further: think of towels and linens taking up 75 percent less space in your senior living home. Mind-blowing, isn't it!

Laundry Pellets

After washing and drying your clothing, the laundry machine in GE's Home 2025 compresses items into pellets and stores them in the machine. When ready to select your outfit, the machine dispenses the items in either pellet form for on-the-go or in revived form to wear now.

This is really not new technology, as much as it would be applied in a more commercialized format that is easier to use. The home shopping networks made mainstream the vacuum-seal bags for clothing that allow you to place your clothes and bedding into a plastic bag and then suck the air out and then seal so that you can get **triple** your storage space. At the very least, senior living should be utilizing this basic technology for their residents to help them manage their space easier.

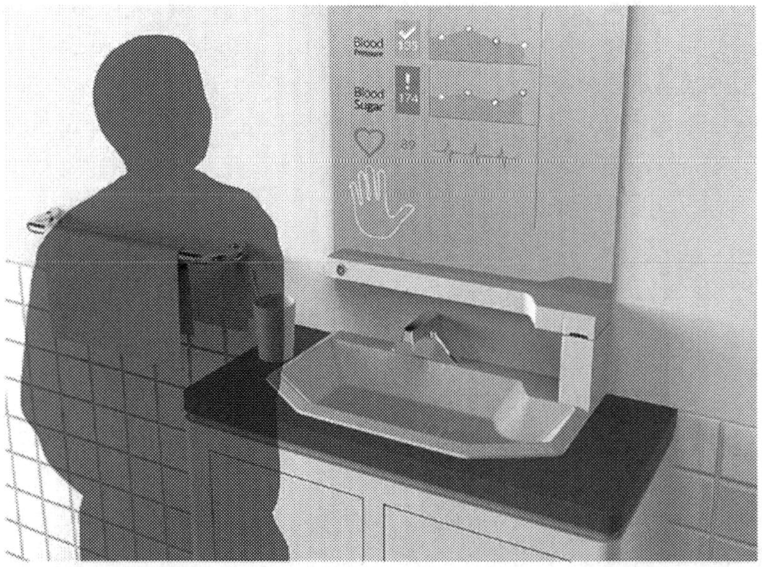

Medical Dispenser

No need to remember what medications to take when living in GE's Home 2025. Just place your hand on the mirror, and the Medical Dispenser reads your vital signs and decides the amount of medication needed for the day. The machine combines, processes and dispenses the medication in liquid form. Medication cartridges can be easily reloaded.

GE's smart faucet envisions that not only would your faucet properly filter your water, but it would also be able to read how hydrated you were and if you needed vitimans, etc., added for optimumn health.

The fact is that with Google currently developing health-sensing technology such as a "diabetic contact lens" that can read your glucose levels through your tears and the XPRIZE's Qualcomm Tricorder contest that has announced its final 10 teams to build the world's first handheld smart phone medical diagnostic tool that can take vitals as well as oyxgen saturation, it's not too hard to imagine that by placing a finger on the facuet sensor, your hydration levels and vitamin and mineral levels could be accessed. Having reservoirs of vitamins and minerals hidden under the sink, you could add them to your drink (no different than adding drops of vanilla to a recipe) along with flavoring if needed or desired.

As a side note: The XPRIZE Qualcomm Tricorder winner will be announced in early 2016 and it looks to change healthcare's future for the better, which will have a ripple affect on senior living.

Taking this concept one step further, GE has reimagined the medicine cabinet to be a combination of the GE imagined faucet and the Qualcomm Tricorder diagnostic tool!

Smart Faucet with Hydration Sensor

The smart faucet in GE's Home 2025 not only dispenses filtered water, but also ice and carbonated water, vitamins and various beverages. Just place your finger on the faucet and the built-in hydration sensor lets you instantly see your hydration level.

Just imagine placing your hand on the mirror so that it recognizes you and takes your vitals, which are then sent to your doctor, or your loved one doing this to see if they need monitoring or if something is not as it should be. The medicine cabinet would then dispense the medications you need in liquid form through the sink. Medication cartridges would be easy to change and medication compliance would be much easier to monitor with constant vitals being taken. No need to be reminded to take this or that; it's all there for you.

Of course, if Roberto dumps his meds down the sink or won't place his hand on the cabinet, you could get a notification of this also.

The best way to imagine this is as a very sophisticated vending machine of sorts. The key will be monitoring of the vitals to ensure medication compliance. Taking too many meds due to forgetting what you have already taken would be a thing of the past.

Paul Hanley feels this is more than plausible and not far out. He states, "As we went through our research we found medicine is gigantic and only growing, and how much medicine these seniors are taking is enormous. Whether they need them or not, the idea was instead of having to remember all that, you don't have to worry as much. It's a little bit of a different paradigm, where the senior says it's OK because I'm going to get up in the morning, I brush my teeth, I'm by my mirror, it dispenses … it's a smart dispenser. It dispenses for that day in something I drink right there instead of sitting there on the counter. My mom has one of those Monday, Wednesday, Friday pill things that never seems to work and she dumps them out all over the place, but this would be taking a different look at the issue." The burden would be placed on the technology vs. the senior, with built-in reminders and safety features to ensure the best results.

Paul's mother is not alone. If you're over 50, you can identify with trying to manage your prescriptions and supplements. It's easy to forget what you took when, especially when so many of these pills look alike, and then we take them out of their containers and unless we are highly organized and disciplined, we overtake or undertake weekly.

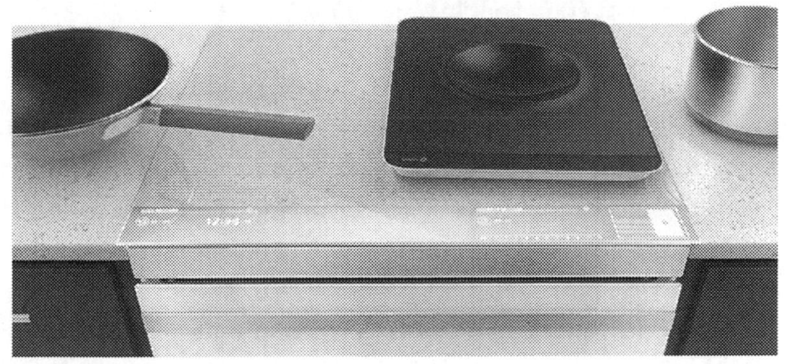

Induction Cooking

Packed in a 27-inch-wide design, the oven in GE's Home 2025 exhibit combines the efficiency of an induction cooktop, Advantium® Speedcook oven, sensor cooking, and a traditional thermal oven into a single unit. The interchangeable and integrated induction accessories allow for unlimited and exciting culinary exploration, as well as more cabinet space when stored in the integrated storage drawer below.

Induction cooking has been around since 1900, but not used by consumers until the 1970s through Westinghouse. Senior living, but especially memory care neighborhoods, have been utilizing induction units for safety and to be able to provide a more home-style dining experience than using steam tables as seen in a skilled nursing facility or cafeteria.

For those not familiar with the basics of an induction cooking unit, I have referenced Wikipedia:

Induction cooking heats a **cooking vessel** **by electrical induction**, instead of by **thermal conduction** from a flame, or an electrical **heating element**. For nearly all models of induction cooktops, a cooking vessel must be made of or contain a **ferromagnetic** metal such as cast iron or stainless steel. Copper, glass and aluminum vessels can be placed on a ferromagnetic interface disk which functions as a conventional hotplate. According to the **U.S. Department of Energy**, the efficiency of energy transfer for an induction cooker is 84%, versus 74% for a smooth-top non-induction electrical unit and an approximate 40% for a gas cooktop.

So why does GE see this as a future technology? Paul Hanely explains that while this is not a new technology, it hasn't caught on. "Partially due to Americans love of cooking with gas; another issue is the lack of understanding and thinking that it's radiating the food. The final is it requires new cookware."

The benefits to induction cooking are immense.

- The obvious benefit is the energy savings.
- Seniors and senior living have the ability to safely cook and then remove the pan or cookware and the surface is safe to touch.
- Homestyle cooking can be performed with the residents in a safe environment that is non-institutional and eliminates the need for steam wells.
- When the induction units are mounted under a granite surface, the space looks just like a regular counter when not in use and therefore can be placed on an island and not hurt the aesthetics of the space.
- Finally, the countertop where induction cooking takes place can be dual purpose. Space is always limited, whether in an apartment or performance kitchen, so being able to utilize the cooktop space as a prep area or serving space when not in use allows the countertop to be more flexible.

With all of these benefits, we still see both existing senior living environments and new-build homes miss the return on investment that induction cooking units provide. It's amazing to see a $17 million project value engineer out the induction units because they will have to buy more expensive pots. Silly, I know, but it happens every day.

Thermal Nightstand

In GE's Home 2025, a thermoelectric nano material allows a user to activate a designated area on the nightstand using their body heat, such as from a finger, and leave a reminder. The thermoelectric nano material also recognizes the temperature difference between the surface and any object you set on it, such as a cup of coffee, and will adjust to keep your beverage warm (or cool).

While I love the idea of smart furniture and Corning Glass, "Day in the Life" video makes a hard case for all of our surfaces becoming interactive computer screens, this nightstand seemed a bit out there. Think of a digital Etch A Sketch that's able to warm or cool your drink and know which one it should do.

GE assures me that this is not so out-of-the-real as both the thermoelectric nanotechnology exists, as well as the heating and cooling technology. It's the combining of these technologies that will take some time.

Being able to leave notes for a loved one or resident to start their day off with a smile or reminder is a wonderful idea. Corning would also have this surface light up and become the alarm clock.

The GE future kitchen has everything from ultra-smart appliances (something out of *The Jetsons*) to the induction cooktop technology that has been used by the public since the 1970's. The key is that the appliances are integrated ino your food delivery system, your smart phone technology, and 3D printing.

If anyone is familiar with PeaPod (mostly in large cities), which started in 1990, it delivers groceries to your door from your online order for a nominal fee. GE envisions this as a step up from this, having the appliances have storage units, similar to a freezer case in the grocery store, that can be stocked from outside the house and accessed from inside the house. The stocking method would be completely secure and medications could be delivered this way as well. You would determine your basics and order them up and then meal-plan, and the appliance would calculate how much you needed of what for your recipes and servings and order the proper amounts. GE envisions you would possibly have a personal micro-garden to grow items that you enjoy growing yourself or are not available at your store.

The appliaces would be a combo refridgerator/oven that would allow you to prep a meal for cooking (plate up), similar to a crock pot, and then start cooking at the approprate time.

They also envision that leftovers could be utilized with a 3D food printer for pet treats. Although the technology is currently available to 3D print meat, they feel it will take some time for people to adapt to this technology.

The focus is on how can we have fresh food without it going bad when you're only cooking for one or two. The anology was given of making tacos for one and the whole head of lettuce would go bad, as the persona only needed a small amount. Seniors tend to reduce cooking due to the issues of cooking for one with leftovers. If we could reduce this and engage cooking with interactive learning, the whole process would become enjoyable again.

In summary, Paul Hanely says, "Looking back, I think that one of the interesting things about this 2025 project was we were grounded in current technology or technology that's either pretty far developed or pretty far in it. We couldn't just say, your kitchen is going to float, or your clothes are going to automatically be clean by solar light, or some new element. Everything had to be grounded in a technology that we either A, know it's worked or B, is right there. I think that gave the designers a little bit more of a challenge, but I think you come out with better ideas when you do that kind of stuff.

"So people aren't like, "Yeah, hovercraft," or whatever it is. I don't get it. This kind of stuff is out there. As I mentioned before, the little T-shirt things that they shoot out at sports games, our laundry concept uses that same technology.

"We are living in the opposite world of you, by the way. That's why these projects are so fun for our designers. Our teams include industrial designers, user interface designers, UX and engineers and researchers and our internal user insight team. It was good to stretch everybody's wings and see what's possible instead of try to make the millionth dryer of the year."

In January of 2015, after this interview, GE unveiled smart phone appliances that let you do everything from preheat your oven to send a warning to you when your clothes have been left in the dryer too long and will wrinkle soon.

While we have discussed many technologies and their benefits, it all comes down to one question: How will technology help you to increase the quality of life of your residents, reduce your risk, and improve your bottom line? Once this is answered, you will be able to confidently move forward with adopting which new technology is right for you and your residents.

We need not be afraid of the future, for the
future will be in our own hands.

— Thomas E. Dewey

Chapter 11

Summary, Bonus and Resource Link

Summary:

The Jetsons aired in 1962 and was set in the future 2062, they gave 100 years for this imagined world to become reality, yet it's only taken 53 years. While some of the technology on the show is not as mainstream today as others (flying cars), we can expect that as the past 10 years have felt like technology was created at light speed, the next 10 will fundamentally change how we all live.

We have discussed technology that can help the blind to see, the deaf to hear and the disabled to walk again. Self-driving vehicles and other mobility devices that increase our freedom and safety. Lighting that has become interactive and aides in our wellbeing as well as the planets. Crowdsourcing and gamification that can help to complete simple tasks to complex tasks by leveraging our connection to others through the internet. Behavior modification through gamification can make you physically and mentally healthier, increase the compliance of your team and residents or save lives. Robots that reduce stress on caregiver's bodies, engage with residents or deliver their medications via drone. While data mining and artificial intelligence run as a part of most of the technologies presented they deserve their individual kudos for diagnosing medical conditions and helping to learn our behavior and patterns so that we are not inundated with information that is not relevant to us. We have been introduced to 3D printing that will change how our goods are delivered to us similar to having a computer and printer in your home. 3D printing will also increase the dignity we can provide for our residents by appealing to the look of food not just the taste without causing a choking issue.

Finally, we looked into the future with GE's 2025 Home and saw how the need for clothing storage could be substantially reduced and furniture and appliances becoming dynamically integrated with technology will help seniors stay in their homes longer.

Some future technologies to think about that we have not covered in this book but have interesting implications are, LED lite highways, a breathalyzer that can detect disease, nanotechnology, virtual reality, We:eX Navigate Jacket and Corning's Day in the Life video which features how important glass will be in the application of technology in our surroundings.

The Future is Here … Which technology will you embrace or adopt?

Bonus:

As part of a group called Abundance 360 that Peter Diamandis runs, I get access to not only the top technology geniuses in the world but also Peter's brain. He is truly one of the great thinkers of our day and freely shares his knowledge with all. While not all of these may apply, I hope you will find it as thought-provoking and exciting as I have. Enjoy!

Peter Diamandis, Founder of the XPRIZE — Top Tech Picks for 2015

1. **Virtual Reality:** Expect a lot more action on the virtual and augmented reality front. 2014 saw the $2B acquisition of Oculus Rift by Facebook. In 2015, we'll see action from companies like Phillip Rosedale's High Fidelity (the successor to Second Life), immersive 3D 360-degree cameras from companies like Immersive Media (the company behind Google's Streetview), Jaunt, and Giroptic. Then there are game changers like Magic Leap (in which Google just invested over $500 million) that are developing technology to "generate images indistinguishable from real objects and then being able to place those images seamlessly into the real world." Oculus, the darling of CES for the past few years, will be showing its latest Crescent Bay prototype and hopefully providing a taste of how its headset will interact with Nimble VR's hand- and finger-tracking inputs. Nine new VR experiences will be premiering at the Sundance Film Festival this year, spanning from artistic, powerful journalistic experiences like Project Syria to full "flying" simulations where you get to "feel" what it would be like for a human to fly.
2. **Mass-market robots:** 2014 saw the acquisition by Google of eight robotics companies. 2015 is going to see the introduction of consumer-friendly robots in a store near you. Companies like SU's Fellow Robots are creating autonomous "employees" called Oshbots that are roaming the floors of Lowe's and helping you find and order items in their store. We'll also see Softbank's Pepper robot make the leap from Japan to enter U.S. retail stores. Pepper uses an emotion engine and computer vision, to detect

smiles, frowns, and surprise, and it uses speech recognition to sense the tone of voice and to detect certain words indicative of strong feelings, like "love" and "hate." The engine then computes a numeric score that quantifies the person's overall emotion as positive or negative to help the store make a sale. At CES, Paris-based start-up Keecker will show off a robot that doubles as a movie projector after raising more than $250,000 for the idea on Kickstarter.

3. **Autonomous vehicles:** In 2015, we will see incredible developments in autonomous vehicle technology. Beyond Google, many major car brands are working on autonomous solutions. At CES, Volkswagen will bring the number of car brands on display into double figures for the first time this year. Companies like Mercedes say they will show off a new self-driving concept car that allows its passengers to face each other. BMW plans to show how one of its cars can be set to park itself via a smartwatch app. And, Tesla, of course, has already demonstrated "autopilot" on its Model D.
4. **Drones everywhere:** 2015 will be a big year for drones. They are getting cheaper, easier to use, more automated, and are now finding more useful and lucrative applications. These "drones" include everything from the $20 toys you can buy at RadioShack to the high-powered $1000+ drones from companies like DJI and the super-simple and powerful Q500 Typhoon. These consumer drones equipped with high quality cameras and autopilot software are military-grade surveillance units now finding application in agriculture, construction and energy applications. Drones get their own section of CES in 2015 with a new "unmanned systems" zone. Wales' Torquing Group could provide one of its highlights with Zano, a Kickstarter-backed quadcopter small enough to fit in your hand but still capable of high-definition video capture.
5. **Wireless power:** "Remember when we had to use wires to charge our devices? Man, that was so 2014." Companies like uBeam, Ossia and others are developing solutions to charge your phones, laptops, wearables, etc. wirelessly as you go about your business. And this isn't a "charging mat" that requires you to set your

phone down … imagine having your phone in your pocket, purse, or backpack, and it will be charging as you walk around the room. Companies are taking different approaches as they develop this technology (uBeam uses ultrasound to transfer energy to piezoelectric receivers, while Ossia has a product called Cota that uses an ISM radio band, similar to Wifi, to transfer energy and data). Look out for a key "interface moment" in 2015 that will take wireless power mainstream.

6. **Data & machine learning:** 2014 saw data and algorithm driven companies like Uber and AirBnb skyrocket. There is gold in your data. And data-driven companies are the most successful exponential organizations around. In 2015, data collection and mining that data will become more turn-key. Platforms like Experfy, for example, allow you to find data scientist who will develop algorithms or machine learning solutions for your business/project. Larger companies can explore partnering with IBM's Watson Ecosystem, which is creating a community of everyone from developers to content providers to collaborate and create the next generation of cognitive apps. Companies built around algorithms, like Enlitic (a company that uses machine learning to detect tumors and make medical imaging diagnostics faster and cheaper), will become much more prevalent and common in 2015.
7. **Large-scale genome sequencing and data mining:** We are at the knee of the curve of human genome sequencing. In 2015, we will see explosive, exponential growth in genomics and longevity research. As the cost of sequencing a single human genome plummets by orders of magnitude (now around $1,000) and the amount of useful information we glean from mining all that data skyrockets. At Human Longevity Inc (HLI), a company I co-founded, we are aiming to sequence 1 million to 5 million full human genomes, microbiomes, metabalomes, proteomes, MRI scans, and more by 2020. We're proud to have Franz Och, formerly the head of Google Translate, as the head of our machine learning team to mine the massive amount of data so that we can learn the secrets to extending the healthy human lifespan by 30-40 years.

8. **Sensor explosion:** In 2015, expect "everything" to be "smart". The combination of sensors and wearables, increased connectivity, new manufacturing methods (like 3D printing), and improved data mining capabilities will create a smart, connected world – where our objects, clothes, appliances, homes, streets, cars, etc., etc. will be constantly communicating with one and other. Soon, there will be trillions of sensors throughout the world. These sensors won't just power smart ovens and sweatshirts -- the same technology will allow companies like Miroculus to create "microRNA detection platform that will constantly diagnose and monitor diseases at the molecular level." Sensors are going to be taking over CES this year. Among the many applications: a shirt that can read your heart-rate by Cityzen Sciences, a device from HealBe that can automatically log how many calories you consume, a garden sprinkler system from Blossom that can decide when to switch on based on weather forecasts, pads for the pantry from SmartQSine that allow you to keep track of how much of your favorite foods are left, a pacifier from Pacifi that sends the baby's temperature to the parent's smartphone, and a new home security system from Myfox with tags you can attach to a door or window that trigger alarms before a break-in is attempted.
9. **Voice-control and "language-independent" interaction:** Using our fingers to operate smartphones/technology was "so 2014". In 2015, we will see significant advances in voice-controlled systems and wider mass market adoption. Think the first steps towards a Jarvis-like interface. Siri, Google Now, Cortana, and other voice control systems are continuing to get better and better – so much so that they are being almost seamlessly integrated into our technology, across platforms. Soon, almost all connected devices will have voice-control capabilities. Companies like Wit.ai are creating their own open-source natural language interfaces for the Internet of Things and for developers to incorporate into their apps, hardware, and platforms. Jarvis-like systems like the Ubi and Jibo, plus IBM's Watson and XBOX One Kinect, already allow natural language interactions and question/answer like commands. Then, Google Translate, Skype Translate, and others

are creating software that allows real-time translation between languages, further eliminating cultural and geographic barriers – the Star Trek universal translator is just around the corner!

10. **3D Printing:** 3D printing will continue to grow rapidly in 2015 as the number of applications increase and as printers, scanners, and CAD modeling software become more accessible, cheaper, and easier to use. 2014 saw the first 3D printed object in space, by SU company, Made-in-Space. 3D Systems continues to innovate around the clock and is releasing a plethora of exciting things in 2015, including 3D printed food and customized chocolates. Three years ago there were just two 3D printing firms at CES. This year there promises to be more than 24.
11. **Bitcoin:** While 2014 was a rough year for Bitcoin (it was ranked the "worst performing currency"), I am optimistic that 2015 will be a better year for the cryptocurrency. Weak currencies and uncertainty in the global economy, emerging smartphone markets in developing countries (billions coming online for the first time), better "interfaces", and more commercial adopters who accept bitcoin as a form of payment will all play a role in a brighter bitcoin future. Finally, it's worth noting that Apple Pay will ultimately teach an entire generation how to navigate life without cash … making the transition to Bitcoin natural and easy.